TAPETUM LUCIDUM

CONSCIOUSNESS

&

THE HOSTAGE RESCUE OF MEDITATION

BARTHOLOMY

TAPETUM LUCIDUM

FOREWORD

Ursache und Wirkung. — „Erklärung“ nennen wir's: aber „Beschreibung“ ist es, was uns vor älteren Stufen der Erkenntniss und Wissenschaft auszeichnet. Wir beschreiben besser, — wir erklären ebenso wenig wie alle Früheren. Wir haben da ein vielfaches Nacheinander aufgedeckt, wo der naive Mensch und Forscher älterer Culturen nur Zweierlei sah, „Ursache“ und „Wirkung“, wie die Rede lautete; wir haben das Bild des Werdens vervollkommnet, aber sind über das Bild, hinter das Bild nicht hinaus gekommen. Die Reihe der „Ursachen“ steht viel vollständiger in jedem Falle vor uns, wir schliessen: diess und das muss erst vorangehen, damit jenes folge, — aber *begriffen* haben wir damit Nichts. Die Qualität, zum Beispiel bei jedem chemischen Werden, erscheint nach wie vor als ein „Wunder“, ebenso jede Fortbewegung; Niemand hat den Stoss „erklärt“. Wie könnten wir auch erklären! Wir operiren mit lauter Dingen, die es nicht giebt, mit Linien, Flächen, Körpern, Atomen, theilbaren Zeiten, theilbaren Räumen —, wie soll Erklärung auch nur möglich sein, wenn wir Alles erst zum *Bilde* machen, zu unserem Bilde! Es ist genug, die Wissenschaft als möglichst getreue Anmenschlichung der Dinge zu betrachten, wir lernen immer genauer uns selber beschreiben, indem wir die Dinge und ihr Nacheinander beschreiben. Ursache und Wirkung: eine solche Zweiheit giebt es wahrscheinlich nie, — in Wahrheit steht ein continuum vor uns, von dem wir ein paar Stücke isoliren; so wie wir eine Bewegung immer nur als isolirte Puncte wahrnehmen, also eigentlich nicht sehen, sondern erschliessen. Die Plötzlichkeit, mit der sich viele Wirkungen abheben, führt uns irre; es ist aber nur eine Plötzlichkeit für uns. Es giebt eine unendliche Menge von Vorgängen in dieser Secunde der Plötzlichkeit, die uns entgehen. Ein Intellect, der

Ursache und Wirkung als continuum, nicht nach unserer Art als willkürliches Zertheilt-und Zerstücktsein, sähe, der den Fluss des Geschehens sähe, — würde den Begriff Ursache und Wirkung verwerfen und alle Bedingtheit leugnen.

Cause and effect. We call it "explanation": but "description" is what distinguishes us from older stages of knowledge and science. We describe better, – we explain just as little as all our predecessors. We have discovered a diverse succession where the naïve man and researcher of older cultures saw only a duality, "cause" and "effect" as they say; we have perfected the image of becoming but haven't gotten over the image, nor behind the image. The series of "causes" in every case stands much more complete before us, and we conclude: this and that must first precede, for this one to follow – but we haven't thereby *comprehended* anything. The qualitative aspect, for example in every chemical process, still appears as a "miracle", just as every locomotion; no one has "explained" impetus. And how could we explain? We operate only with things that don't exist, with lines, surfaces, bodies, atoms, divisible times, divisible spaces – how should explanation be possible at all, when we first make everything into *image*, to our image! It is enough to view science as the most faithful anthropomorphizing of things, we learn to describe ourselves ever more precisely, while we describe things and their succession. Cause and effect: such a duality probably never exists, – in truth a continuum stands before us, out of which we isolate a few pieces; just as we always perceive a movement as isolated points, and thus don't see, but conclude. The suddenness with which many effects arise, misleads us; it is only a suddenness for us. There is an infinite quantity of processes in this sudden second which elude us. An intellect which saw cause and effect as continuum, and not as arbitrary division and dismemberment as we do, which saw the stream of events, – would reject the concept of cause and effect and deny all conditionality.

Nietzsche, *Die Fröhliche Wissenschaft*, §112

- 0 -

The time is ripe for the development of our own language for what we do. What is meditation? Why is it effective? Why is it important? These questions can be answered in novel ways that satisfy our accustomed taste for a reductionist approach grounded in physiology and evolutionary history, while not only leaving room for the mystery and wonder of incremental awakening, but perhaps even enhancing it in ways the priestly fables can no longer accomplish. We are the aberration of terrestrial self-organizing matter capable of exploiting the aberration of excess apperceptive consciousness to achieve profound psychosomatic modulation: if you cannot find that story more thrilling than tales of *samsara* and the promise of "the extinction of desire", I cannot help you.

- 1 -

I have hesitated until now to give a theoretical account of the practice of meditation and its effect on consciousness, despite how popular it might prove to be. I have felt that any such theory would occlude and preclude the actual practice – for which no amount of cheap talk can substitute. However, I have seen that students struggle to make sense of their experience, and that too many promising students are led astray by the seductions of the priests, who are not hindered by the inconvenience of an intellectual conscience.

- 2 -

That this work on consciousness and meditation is another opportunity to demonstrate our *Umwertung*: that what we value so much about ourselves – our intelligence, our sapience, our self-reflection – is not itself a source of value but the recipient. That it is the foundations of terrestrial life which yield value. That it is everything we take for granted in the human body that makes us intelligent and wise, if we are so at all. That everything conscious and overt and so celebrated in modernity, is shallow and recent and trivial in comparison to the enormous unconscious intelligence upon which we draw, with every breath.

- 3 -

What follows is a phenomenological description of consciousness in the language of nonlinear systems, topology, and information science: wherever such language is unfamiliar, the reader is encouraged to focus on the ethology and anthropology – it's *observation* and *experience* I'm attempting to describe, not the theory itself. Here we want a description of how and *why* consciousness exists at all, what its organic function could be, how it might come about in the long history of neural development, and what its use might be to us today: none of this should be taken for granted. We should never assume that "consciousness", whatever it's supposed to mean, is a self-evident good-in-itself: no matter the topic, "self-evident" is always a sign of a refusal to see, a refusal to learn, a refusal to discuss.

It's worth pointing out as I have elsewhere, that in any discussion of "apperception" we're necessarily talking about *perception of perception of perception*: if we aren't capable of perceiving ourselves perceive perception, then we have no right to talk about it, and remain just as unconscious about this as we are about so many things. We should ask whether such a thing is possible – or merely another piece of ape vanity. To articulate what is unconscious requires a kind of peripheral vision, an ability to watch without disturbing, the ability to discover traces in the aftermath, to follow consequences upstream in the psychohistory: it's the *dectective's reasoning* in both psychoanalysis and meditation that I draw from, when I describe what it means to "become conscious".

- 4 -

Like so many smart boys of my generation, I'm a former software engineer. This part of my education has probably shaped my approach to "spirituality" more than I can appreciate at close range: what is it to be an engineer but a kind of halfmad tinkerer in many sciences at once? Not only computation and algorithmic design, but information theory, piecemeal statistics, and especially *the fine art of scalability* is what the experienced code monkey must learn...

An engineer is to a scientist what a touring musician is to a chamber orchestra: half the job is just keeping your shit organized and yourself sane while you try to introduce a little craft and artistry into the messy business.

- 5 -

I am not a Buddhist. I don't possess a borrowed Japanese title, nor did a Pakistani pandit bestow a "Gurudev" upon my willing brow. I belong to no order of priests. I belong to no toothy grinning cult. I have been among all these types and examined each in turn: in every case I learned something, but declined to stay. I respect aspects of many traditions but I choose to walk a different path. I have had many teachers but encountered few masters.

- 6 -

When you read this work, you're reading a lifelong student of the humanities coming to grips with mathematics. You'll notice that I study math, physics, and computer science very differently than the 21st century head-in-the-sand specialist with his toppling stack of inherited terminology. Bad teachers can't handle the genuine question, because they only have *answers*: always keep this in mind.

When I don't understand, I turn to my strengths: history, etymology, philology, philosophy, anthropology, and psychoanalysis. I study mathematics like a historian of ideas. This is what it means to me, to be a *homme de lettres*: to resist being overawed by the illusion of certainty and completeness in any field, to study everything as a genealogy of cultural prejudice, to be a student of humanity especially where there is so little humanity.

- 7 -

I've said this elsewhere but it's worth repeating: not only have we yet to understand the nervous system, and not only have we yet to formulate a consistent definition of sentience and consciousness, we still don't know *what neural information is* – neither how it's stored, nor transmitted, nor translated.

For example, most researchers like to ignore that a large portion of the neural population isn't composed of integrators that sum their inputs, but resonators that respond to a narrow frequency band. And in a related vein, that stochastic resonance has already been demonstrated to play a role not only in hearing but in tactile resolution...

The state of neuroscience is something like astronomy in the late Renaissance: already losing touch with the descriptive power of the ancient Ptolemaic system without having established any of the predictive basis of the Copernican. Our psychology grows weaker by the day, while our Bayesian tautologies multiply: hardly anyone has noticed that as our commitment to premature quantization increases, our sense for what we're studying diminishes. What's the point of consulting immediate bodily experience if your body is a miserable hindrance to greater production of intellectual ware? What is psychology but the taxonomy of private resentment and impeding neurosis – something increasingly considered as worthy of real science as *landfill management* is to ecology. The neuroscientist today looks at the human being and sees a golden brain on a shabby stick, a prisoner of inertial dampening, a damsel in a dirty tower, a kind of sick joke of evolutionary whim.

That I expect to be believed, when I talk of "neuromorphic experience" as something both perceptible in every moment and hidden by virtue of its efficiency, as something salient in meditative discipline and psychedelic excursion, is another example of why originality is such a gambit: every novel term of discourse is a bargain made with your audience, in which you promise something you cannot demonstrate until they learn a language no one else seems to be speaking. So until these terms and these approaches gather enough social currency – and it's happening faster than I would have thought – we have to content ourselves with lonely vistas: but it was precisely the lonely vista that inspired this study, so on the other hand by the time these terms reach social acceptance we will have likely moved on to fresher waters.

- 8 -

What I hope to demonstrate in this little book, is that rigorous rationality and profound aesthetic appraisal of mystery, are not only commensurate but *mutually reinforcing*: the more articulate we become, the more accurately we carve the negative space of ignorance. The more committed we are to a strictly limited but reliable method, the more we can afford expensive admissions of uncertainty. Mystery is not free, but a luxury of clarity.

PART I:
A HOSTAGE RESCUE

A LIVING HEURISTIC

Mein ganzes inneres Wirken erwies sich als eine lebendige Heuristik, welche, eine unbekannte geahnete Regel anerkennend, solche in der Außenwelt zu finden und — in die Außenwelt einzuführen trachtet.

My whole inner working showed itself like a living heuristic, which while obeying an unknown intuited rule, seeks to find the same in the outer world – and to introduce it into that outer world.

Goethe, *Maximen und Reflexionen*

\- 9 -

I see young people hungering for the union of science and spirit: for the right to profound communicable experience with a good conscience, a sense that we can gain back the core human adventure without resorting to delusion and grandiose error. That if spiritual development has a basis in anything but pathology, that it must therefore yield to scientific investigation. That the problem can be framed in terms of unknown planes of causality which still elude our analytic and anatomical methods. There are many traditional disciplines which we know work, but we don't know why: the old attitude didn't need to know why. What we are increasingly unwilling to tolerate any longer, is the assumption that if we don't understand it, *it must not exist*: in fact, this is *highly unscientific*. We are not somehow more intelligent than our ancestors: in many ways the situation is reversed. The accumulated evidence says: the more ancient the practice, the more likely it is to be valid and worth taking seriously. For example, meditative discipline works: not only does it "reduce stress", it has the power to permanently reshape a psychosomatic topology.

In this domain, what I call "cargo cult scientificality" must be defended against vigorously: we must not give in to gratifying postures of certainty we have not earned – most science is popularly presented in the priestly garb of pseudoscience, and most of the spiritual schemes around us now take this very same pseudoscientific posturing as prerequisite. One hears the scientists lobbing the charge "pseudoscience" more often than before: perhaps this is because they feel their largely unrewarding toil has earned them a sole right to *unscientific attitudes of certainty*?

- 10 -

Somehow we have to give room sufficient for innate and superior knowledge to develop, without permitting the perennial arsehole, nor the insufferable hippy, nor the neon charlatan.

Somehow we have to avoid becoming mired in the bloodfeuds of epistemology, while providing a sufficiently cogent and satisfying theory of this kind of knowledge: we have to quench our thirst for scientific righteousness, while feeding our hunger for spiritual relevance. We have to find a way of knowing what we need to know, without lying to ourselves about how we know it.

We would like to develop a theory of method alongside our theory of result, in order to refine and accelerate future inquiry. With our newly hybridized scientific techniques, we would like to do more than justify ourselves and seem impossibly hip.

We would like to avoid constructing elaborate illusions of having arrived where we are through established practice. We would like to be honest about having stumbled and fumbled our way into clarity: but a theory of productive stumbling and fumbling is not therefore precluded.

- 11 -

Are we doing more than negotiating productive compromise between the forces that would possess our bodies? Building bridges of reconciliation between seemingly incompatible modalities of the past? Are we merely greedy for authority? In our haste to unite the reductive powers of science with the holistic redemptions of spirituality, are we doing more than borrowing and abusing technical vocabulary, or are we beginning to accurately and evocatively describe subjective experience? Isn't *precise description* the bedrock of good science? And isn't the ability to formulate ever more articulate questions the jackhammer leading to this bedrock? And how do we become articulate in our inquiry, but by borrowing and inventing technical vocabulary?

- 12 -

The purely *descriptive* phase of a given field requires more scientific acumen than the later reductionist experimental stages. Does anyone know this anymore? Does anyone possess an intellectual conscience alongside a fruitful imagination? This is why good anthropology is so hard to come by: most of them are not psychically malleable enough for an immersion, nor are they disciplined enough to describe ambiguous valence: they have neither realized subjectivity nor rigor – somewhat like comparing a Rothko to a Dürer.

This is also the case with almost all modern psychology, since psychology should properly be treated like an anthropological adventure. Most of civilized humanity, when in the presence of a novel experience don't become dumbfounded, they become "smartfounded" – they erase the novelty with certainty. They *lack the character* to assimilate new information: it requires humility, patience, and assurance to look into the unknown and perceive novelty. Mere intelligence among human beings is cheap and ubiquitous; character is expensive and rare.

- 13 -

Despite all the abuse it's received, the concept "soul" is not entirely dispensable. Higher-order emergent patterning, the mysterious recursions of complexity, and everything tantalizing about synergetics, speaks of "soul", doesn't it? What's initially so easy to regard as mere epiphenomena, sheen, byproduct, at some hypercritical moment, seems to reach out of an ether and pluck a sympathetic string in the plenipotent present. At this time our notion of unidirectional agency is challenged: our conceptualizations of causality have been naïve and a little blockheaded. The reductionist perspective is a useful discipline and the gleaming scalpel of Western science – but it blinds us to circularity, dependent co-arising, and more subtle causation models. Sometimes the whole guides the parts. Sometimes our dreams speak with a clarity and prophecy that cannot be denied. Sometimes we just know things we can't explain.

But the first thing I need to say after having articulated this, is that this knowledge and this practice is *not for everyone*. Most human creatures are so far gone down the road of self-deception, mendacious hypochondria, and entrenched wishfulness in relation to everything "holistic" – they may name it morality, idealism, or increasingly just "self-care" – that even the word "intuition" entails far too much trust for apenature not to abuse it.

Reductionism is not a necessarily *more true* vision of the world, it is a *method* for dealing with the human proclivity to lie to itself. One of the most common blunders I find among the low-level wouldbe scientists with which we are surrounded, is *to confuse efficacious method with objective modeling*. We have become extremely adept at analyzing into constituent parts, we have yet to begin to learn to assemble into wholes.

It's telling that the initially clunky assumptions of thermodynamics in postulating a bounded system, form the basis of the only scientific account of a whole. Thermodynamics was invented for the analysis of machines – specifically steam and combustion engines. The world as machine, the body as machine, life as machination: this familiar Cartesian perspective makes us comfortable enough to draw a line around an assemblage of parts and ask, how does the whole operate? Or how does complexity within limitations affect the behavior of the system as a whole?

But Descartes was not the first modern philosopher, he was the last of the scholastics. The machine does not threaten the foundations of scholastic logic: we retain *deduction* as the leading actor, while inductive inference remains in the background and at the edges – as mere "application" and the preliminary stages. This obscures the fact that everything deductive logic accomplishes, is *mere tautology*. Most "proofs" are arrived at *after the fact*. It is in the midst of inductive inference that everything essential to science happens – that is, in the *margins*: hypothesizing, synthesizing, realizing, guessing, *understanding*. Syllogistic logic still struts around as crown prince, but it is inductive inference that quietly knits the whole endeavor of science together.

- 14 -

In animo volvere. I've tried employing the term "cognitive science" to describe what I do, but there's something about the word "cognitive" that leaves a foul aftertaste. Those who use the word "cognition" do so to explicitly exclude unconscious valence – *intentionally and unconsciously*. You see I cannot take a single step without violating the limits of everything that's currently ascendant in psychology, anthropology, neuroscience, and the study of intelligence. "Cognitive assassin" might be more apt. But as a scholar and stubborn philologist, I'd rather revive and baptize venerable terms: there's nothing wrong with the term *psychologia* nor this "story of the soul". And even the word "soul" has etymological hues I can live with: tracing the Old English *sawol* gives us the clue of the Greek αἰόλος, meaning "quick", and we arrive at the Proto-Indo-European **u̯el-**, meaning "to turn". This leads us to familiar words like "volume", or "the waltz": to have a soul is to keep the wheel turning. To be alive is to move. This is the same sense in which *psyche* was used, along with *animus*, 气, and a whole pantheon of ancient ideas revolving around breath, motion, "quickening". Psyche is a measure of vitality, and therefore psychology is "the story of being alive". – And what lives embraces a great deal more than "cognition". The problem with the term "cognitive science" is that translated from the Latin, it comes out to mean "knowledge of knowledge". And this is precisely what its advocates want: a knowledge only of that which is known. *A science of the conscious*. And this they achieve, and thus everything produced under its aegis is shallow, brimming with worthless taxonomies and ad hoc abstractions which have nothing to do with personal experience, nor sustained observation, nor the deeper history of psychology. The "cognitive scientist" feels justified in reinventing the field from naïve hearsay and willful ignorance. The most successul in this field manufacture consoling tabulations, quarantining sources of anxiety with prettily painted picket fences and green grass on *both sides*, so that everyone gets to feel humane, included, and safe from *the It* which is namelessly excluded. If I were to recast my work alongside theirs, I would rather it be called the "incognitive science", the knowledge of unknowledge, the story of the repressed.

- 15 -

> [...] theorems are stated as precisely as possible; and though they are discovered by intuition, they are demonstrated by logic alone. This habit of mind is called by mathematicians "rigor".
>
> R. Trudeau, *Introduction to Graph Theory*, §2.3

What's the difference between intuition and logic? Intuition is a result, logic is sequential. Intuition is the shorthand of perceptual approximation, logic is the set of rules governing discourse: perception versus narration, understanding versus explanation.

Logic trumps intuition in the paradox: but only to the degree that discursive logic had misplaced the intuitive assertion. Generally paradox arises because of *heterogeneity*: an intuition is generally not "wrong", only *misunderstood* by stupider elements, often with ulterior motives of their own.

- 16 -

What's the guiding principle?

To explore the threads of neuroscience, looking for our inspiration as meditation practitioners. To provide a guiding light from the much-neglected other end of the project of *scientia*: that of the *humanities*. To act like *whole* human beings for once, and unite for ourselves our accumulated personal experience with likely hypothesis and rigorous method. To fill the enormous and widening gap in neuroscience: that of a viable and *deep psychology*. To speak for those with more than a passing familiarity with meditative discipline, altered consciousness, psychedelia, the abyss of empathy, a life lived in constant emotional evolution and the formation of character: my task is to speak for those who know much more than they yet know, to represent the deep wells of unconscious competence which animates the human creature, to articulate that psychological acumen which all of us already possess.

- 17 -

How do we talk about meditation from the most inspiring perspective? *As refinement of unconscious intelligence*, as hierarchical remobilization, as alignment of prioritization algorithms and asynchronous function. Neurochemical release is not tantalizing enough. Correlation with regions of white and grey matter is not tantalizing enough. After many years of rumination and circumambulation, we feel convinced – not merely because of how gratifying it might be – that the answer to what meditation is and can be, is to be found among the principles of information theory, signal transduction, topology, and even some of the peripheral theory of artificial intelligence.

21st century folk like to think of themselves as embodied computers. As long as the metaphors remain infused with the assumptions of von Neumann architecture, this is patently false: but if we're careful about *the definition of computation*, the opportunity reappears. For example, can we demonstrate that meditation is like a *defragmentation routine* of a neural topology at criticality?

- 18 -

There *are* gems here for those of us looking to refine our spiritual disciplines: there is more than intellectual candy and short-lived gratification schemes. We will step gingerly around the populist indulgence and moral posturing surrounding words like "consciousness" and "intelligence": these are still recoverable terms which can and *should* be defined rigorously.

Our criterion is that whatever models we employ and whatever metaphors we propagate, they must prove themselves useful for elucidating our *genuine experience*: the guiding principle is that every theory should *reveal more ambiguity* than suppress uncertainty, it should generate more curiosity and passionate questions rather than smug tautological dead-on-arrival endgame-perspectives – as the vast majority of philosophizing seeks to do. You, dear reader, should come away from our conversations inspired and emboldened, remembering the past and imagining the future, a little more prepared to be honest with yourself and articulate with your surroundings. Our work here is an extension of what we want to bring to everything: clarity, appreciation, a falling through the floor, a love affair, a belly laugh, a roll in the mud, a long rambling walk through a mired city and a blossoming wilderness in the cracks and beyond.

- 19 -

One of the primary sources of confusion in neuroscience and cognitive psychology, is the tangled briar surrounding the definition of consciousness, coupled with the assumption that fully realized sentience belongs to human beings alone.

They confuse at least 3 things:

¬ *Narrative consciousness*, which is conditioned by symbolic capacity. Contrary to almost everyone's assumptions, this is *not* restricted to human beings because symbolic thinking is not uniquely human: without this caveat we fail to understand its essence, because we continue to ascribe extraorganic origins to something which is profoundly organic.

¬ *Apperceptive consciousness*, which is the perception of perception. This is even less uniquely human.

¬ *Sentience*, defined as a functional sensorium with an origin and navigable space. A sense of being-in-a-world, complete with time and space domains, instinctual urgencies, rivals, rituals, crescendos and transitions: all this is common to all animal life, as any honest ethology will testify.

We ask these questions where almost no one asks them: What is the evolutionary value of the nervous system? Why develop it to such expensive extremes? What is the adaptive value of sentience? What does it mean to "feel"?

- 20 -

The purpose of developing a novel scientific language for meditative practice, is to wash it clean of the *priestly stink*. For centuries now, the postaxial priests have claimed sole right to all serious meditative discipline: among the urbanite western Buddhists this is now most obvious, but previously the Christians and Muslims also participated in seeming to own everything "spiritual". They called it "contemplation of Christ" and barred all other forms of prayer; they called it "devotion to Allah" and barred all other forms of ecstatic song.

The war for the human soul is not decided. Buddhism in name does not represent a truly serious threat to the future of spirituality, but Buddhist-Christian *morality* very much does. The goal therefore is to develop an *amoral spiritual discipline*, one which seeks to exploit civilizational maladaptation in order to explore human potential: we do not know what we could become, in the midst of so much error and illness. One thing is certain: the absurdity of hypermodernity may generally crush the human spirit, but it also produces beautiful monsters with unknown powers. This work seeks to be a prelude in the history of the great experiment: *what shall we become?*

\- 21 -

The priests were the first nihilists: overexposure to the fading rituals of the gods, proximity to the fallout of multilingual literacy and cosmopolitanism, and every usual incentive of the charlatan pushed them to this extreme in the early Axial Age – perhaps even before the merchants and pirates got there with worldweariness...

\- 22 -

Meditation has been employed as just another means of cultural castration: a *neurochemical* castration and hypoxia; a hypnosis of the instincts of violence.

- 23 -

The hostage rescue of meditation: the rescue of "spirit", the rescue of an ideal – but do we want an "ideal"? Or do we want merely the discipline, the attitude of excellence, and not the idiotic worship of the impossible?

I teach meditation as a health regimen. So how do we dispel the priestly stink? By regaining *the ancient conception of health*: the dry Hippocratic empiricism, the twisting snakes of the rod of Asclepius, the yogic "path of kundalini", and our own most reductionistic account of neuronal topology *can* be reconciled.

- 24 -

But in the other aisle, sit the cloistered bores who claim sole right to the privileges of knowledge: the academic scientists. They too are an obstacle: they do not want us to succeed.

By no means do they want meditative practice, psychoanalysis, and psychedelia to yield special insight: they refuse the witness of direct bodily experience, because they fear and loathe their bodies. Despite the fact that there is no other criterion of truth but direct experience, despite the fact that the spirit of science is nothing less than the ability to *learn from experience*, academic scientists no longer cultivate observation nor gather evocative experience. They tweak the dials of machines they lack the engineering discipline and imagination to have invented. They fabricate findings. They exaggerate the importance of their own work. They secure funding and fight for scraps. They are not fundamentally interested in truth: this you must internalize, whenever you are forced to go begging among them for knowledge.

- 25 -

Neuroscience has largely gone astray and does little more than put on pompous airs. It has followed the paths laid by Emil Kraepelin, Santiago Ramón y Cajal, and B.F. Skinner: the path of brutality, the path of willful stupidity, the path of complicity with civilization's worst agencies of punitive normalcy.

- 26 -

What you need to keep in mind about academic science at all times: they are not addressing you as the willing student, nor are they directly addressing each other as colleagues, and they are certainly not speaking to the topic itself. They are speaking primarily to those who control access to their funding.

¬ What are they concerned with? Securing and increasing their status.

¬ How do they secure and increase their status? By justifying their existence to those who control funding.

¬ How do they justify their existence? By appealing to the fundamental concern of the university: *prestige*, which leads to funding.

This is why all the sciences in which there is considerable competition and thus much to gain, such as neuroscience and artificial intelligence, are constantly overstating their knowledge and promising great results in the near future. They must always give the impression that they have recently discovered a vital clue leading to even more impressive results in the next few years: look carefully at nearly any neuroscientific literature from 1910 to 2010 to see this play out at ridiculous scale. In other words, they cannot afford the proper attitude of science: *the admission of ignorance*. Almost nowhere do they state what they don't know: the clever student must deduce this for himself. Almost nowhere do they ask simple rigorous questions. Almost nowhere do they take the time to refine their own observational powers and the clarity of their thinking. To do so, would risk giving the impression that they're not making progress – or at best, only producing trivial and redundant elaborations of established fact. About 99% of all academic research is trivial and redundant. They are in such haste to produce impressive answers, they do not take the time to formulate real questions. The formulation of incisive questions is the signature of genuine education, not the hoarding of answers.

- 27 -

Chinese and Ayurvedic medicine serve as a warning: we must not allow a medieval thirst for crystallinity to distract us from an elegant and articulate pedagogy. A psychologist in my sense, or a "spiritual practitioner" if you prefer polluted terms, should be trained like the *naturalist* was not so long ago: a cultivation of the art of field observation means learning to sketch the subject with delicacy and affection, to form working hypotheses in the midst of intense personal experience, to keep one's head even as one risks one's neck. This was once considered the status quo in the training of a worthwhile biologist. Despite the technocratic amplifications afforded by the age of the transistor, I believe science as *formative discipline* peaked in the 19th century – but not before producing such wunderkinds as Darwin, Maxwell, Freud, and the brief blossoming of anthropology. If what I'm outlining has any chance of improving our spiritual life, then it aims not at the achievement of a much-wished-for "objectivity", but at the formation of our intellect into a useful and precise instrument. Every instrument has bias and occludes more than it reveals: hence I don't seek a synthesis of my teachers, but refer to them like planetary influences and distinct astrological "stations". Any teacher worth his salt is almost *incommensurate* with every other. They should be treated and honored like local gods: in their *vicinity* we do them homage. It's hoped it will be understood I'm interested not in ostentatious namedropping, but that with every name I have a verifiable *personal* relationship: I have felt and seen something in their presence no one else could make me feel and see. This is why an intellectual life in my style should feel like an unforgettable party, wherein you drift from a life-changing conversation to ecstatic dance to introspection and back again. Those who know how to party don't merely goof off, they *make medicine*: the human being has *constant need of medicine* – that's what the Plains Indians are trying to tell us with that phrase. This is generally why I refuse to see spiritual life as some kind grandiose "entheogenic" enterprise: we're not becoming divine – our first job is to become *fully human*. Spiritual life is largely *remedial*, recuperative, and convalescent. It's not until we understand how badly we need medicine, that we begin to develop the special powers of the convalescent – almost everything worthwhile in spiritual practice is first generated as a psychosomatic scar. What sometimes seems to make the crises of modernity seem like peak human realization, is its potential for bizarre hypertrophic speciation: chronic psychosomatic disease largely crushes souls under its wheels like the juggernauts of old, but occasionally a fragment of some other ore is wedged just so and fires off in an improbable direction. We are such discarded pebbles. We are the ejecta of a catastrophe: may we one day raise the eyebrow of a beachside strolling geologist of the human soul, who can read past the wellworn surface into the same compressive and tensile forces with which he must still contend.

A LIVING HEURISTIC

THE BIG WHY

- 28 -

More than any other 21st century endeavor, people look to neuroscience hoping to find an answer to the *big why*. But this it cannot provide: science does not yield answers to questions of purpose and meaning. But nor does it preclude them: it is one of the hallmarks of a creeping unconscious nihilism to covertly employ the attitudes of science to foreclose the possibilities of unforeseen growth, perspectival shift, radical revaluation, the undiscovered country of our ancestral past. We do not yet know what we are: neither what we have been, nor what we are now, nor what we're becoming. We are magical apes, shapeshifters, blowhards, artisans, wizards, liars and lovers: that much can be said for sure. But a definitive answer to what it all could mean, this brief window, this vanishing vista, this trainwreck of civilization, this continually cresting history, this cascade of neural activity we seem doomed to only amplify – that we do not have.

The discipline of meditation will take you a step closer: in that much the ascetics have always been right. But we're here to correct those traditions: because meditation, like all esoteric spiritual practice, positively does *not* yield definitive final answers to anything at all. No nirvana, no final revelation, no samadhi, no high that lasts, no insight that makes you immune to confusion, no feeling that cannot fade. What it does, is change your expectations. It lightens, it moves, it teaches us to ask better questions. It teaches receptivity, gentleness, the power of quiet tender passion. I said to a student recently: "Fear is an excellent meditative fuel." And the astonishing answer I received was: "Sometimes fear rises up from my belly, crests and crashes down, becoming ambiguous euphoria." This is the strange power we're cultivating: to unbind emotional valence through the power of ambivalence, the power of ignorance. Not to know, when everyone tries so hard to seem to know. To ask, "What is this? Is this feeling what I think it is? Is this thought so bad? Is this fantasy so good?" Just a little room, a few milliseconds of hesitation between stimulus and habitual response, is all that's required for massive systemic change. Once you begin growing like this, you cannot stop: that too, is a kind of curse disguised as a blessing disguised as a curse. There is tremendous inertia on both sides of mediocrity: it's a steep slope into esoteric neuromorphic calibration routines, for those of us unwise enough to take the challenge seriously. *Unwise*, in the sense that everyone who chooses stultification and stasis is so *streetwise*: to get the rewards of exotic self-awareness requires that we become profoundly foolish. We are fools for growth, fools for receptivity, fools for insight, fools for a deepening passion of living on the edge of breakdown.

Critical emergence as my prime metaphor for the leverage of meditation and the meaning of apperceptive consciousness, is not chosen lightly: there is a reason the ancient practices require stress, hardship, endurance, *crisis*. Biological systems do not function at their peak self-organizing basins of attraction until they find their way there through paths of least resistance: which means all easier paths are no longer possible due to the volume of dissipative flow – which means, *shit gets tough*. Which is why we insist on naming what we do, meditative *discipline*. This is the meaning of ascetic practice itself: some deprivation is always necessary to generate criticality. You have to want to find your way out of a trap you set for yourself: that is the cruel magic of spiritual discipline, and why so much of what we do has been confused with neurotic self-punishment and priestly schemes. But part of us always knows precisely what we're doing, as though we've already been there before. Such is all growth, no? The genome knows where it's going when a body grows: why shouldn't the "spirit"? And if we take that to mean, the critical emergent moments along the trajectory of a self-organizing informative process, as evoked by this neuromorphic chemical bath which speaks to you now, why shouldn't those moments be "already known" in the sense that all harmonics are anticipated and lightly visited by the basal tone? Complex systems don't precisely evolve in a linear stepwise fashion, but rather jitter and switch between their many modes – and the more energetic the system, the more it's already been to those moments: which means, you have already visited many of your superior growth factors, you have briefly been much wiser than you normally are. Perhaps in dreams, perhaps in your most doubtful monologue, perhaps in the midst of loss, perhaps in a fragmented repressed zone you dare not speak of: most of what is most valuable about almost everyone I meet, lies hidden and covered with shame. Most of us are *terrified of breakdown*: precisely where you will find that version of yourself which, truly, the rest of you *should* be afraid of.

THE BAD PSYCHOLOGIZING OF NEUROSCIENCE

\- 29 -

Every theory is composed of metaphor. And conversely, every metaphor is a little theory. Beware therefore, the slipped mickey of suspect metaphor! Punitive normalcy only succeeds to the degree that suspect metaphor succeeds.

\- 30 -

The overwhelming majority of what passes for psychology is loaded with unexamined, clumsy, reifying metaphor. It is language which *forecloses by design*. It is thought which precludes thought. It is repressive schemata disguised as inquiry. It is morally charged normative reinforcement emboldened by institutional support. It is the fine art of anticipating the mythopoetic needs of dominant power trajectories and translating their justification-of-rule into a corresponding justification-of-subjection. A "scientific psychologist" is the corollary to the "political scientist": the first *flattens* the subjected, the second *flatters* the subjector. Has it struck no one else that modern psychology transforms inexhaustible subtlety into threatening blunt objects, while political science transforms the threat of violence into delicately woven rationality?

\- 31 -

Consider for example, the highly influential *premature scientificality* of Wilhelm Wundt and William James: which is merely the translation of late-scholastic psychophysical parallelism into 19th century scientific verbiage – "consciousness" in lieu of "soul". The metaphor has not fundamentally changed. Most 21st century psychology has proceeded no further – and in fact regressed from the accomplishments these two uninsightful bores: at least 19th century scientists were consistent.

Why do people say "subconscious" and "subliminal"? Because of *unconscious spatial metaphor*: most of humanity still imagines a little man in the driver's seat of the brain, where everything below his seat is as messy, unknowable, and frightening as the innards of a car is to most drivers.

- 32 -

Why study topology? Because psychologists are always employing *spatial metaphor* without knowing it's spatial, nor acknowledging it's a metaphor, nor once studying the properties of space.

Psychology as the study of *virtual space*: there appears to be no better way to describe subjective experience.

- 33 -

The "gradient vector field" is one of those ascendant metaphors of the 21st century driven by the increasing familiarity with computer science: but it's true that we learn the shape of our neural experience only by traversing it, and this traversal alters its shape. Everything is *impelled*: thus we "flow" along various trajectories and remain relatively blind to the future, although our neural experience is also a summation of many such trajectories operating asynchronously. As the experienced practitioner knows, every breath is a traversal with many adjacent paths but only one optimal minimum.

- 34 -

My goal is not necessarily to solve any theoretical problem in neuroscience, but to generate better metaphors in psychology: to develop a metaphorical language drawn from promising mathematical disciplines. We want metaphors which are less likely to lead us astray, because they are less replete with unexamined metaphysical assumption – as almost all psychology still is. I suspect that in the coming years I will turn out to have been only one example among several. We cannot help but harbor an ambition to provoke existing neuroscience into more fruitful paths: I hope to inspire mathematicians to meditate as much as meditators to learn to love math.

- 35 -

The lack of a proper mathematical grounding in a liberal arts education in our time is shameful: the fact that I was not taught geometry from Euclid's *Elements*, for example. A taste for pure mathematics was one of the pillars of the ancient notion of *disciplina*: pure mathematics should be offered to the poetically gifted child like a rare jewel.

- 36 -

For example, there is the possibility that "isomorphism" in graph theory should be applied to the *semantic field*: a conceptual cluster is isomorphic to another, if the vertices and edges remain intact despite deformations due to aspect and arrangement. In other words, a *metaphorical* relationship is an *isomorphic* relationship: and narrative consciousness is driven almost entirely by the glorious contortions of metaphor. No matter how exotic the story, it's a matter of human instinct to place our "I" in the position of protagonist, and seek metaphorical lessons from the narration.

- 37 -

This work is not exactly a critique of existing neuroscience itself: for example, the components of chemistry and anatomy are unassailable. What we're attacking is the underlying set of *psychological assumptions* which permeate the field and make it stupid: they make constant recourse to psychological concepts they have not once analyzed. A neuroscientist is something between a chemist, an electrical engineer, and a mortician: I have yet to encounter one of them in all their vast literature with any serious psychological acumen. It's telling that the most refreshing perspective I've found in this study, comes from Rodney Brooks, a researcher in artificial intelligence originally trained in pure mathematics. These people don't take their own experience seriously: they have not learned to *observe*, they don't know the value of *case studies*, nor are they the least educated in the past classics. Their research is entirely dependent upon expensive and highly technical implementations, which in the end yield *less than nothing*: an fMRI reveals a little about anatomy, but in absence of any psychological understanding you have learned less than nothing, because you have generated a red herring to distract your colleagues, your students, and *yourself* from your true ignorance of what you're studying.

- 38 -

The question that faces us in neuroscience, is how a continuous series of ionically induced voltage oscillations could be processed to produce such precise behavior in space and time. Does the secret lie in the so-called neural code, does it lie in the more sophisticated "preferential charge path computation" of neuroelectrodynamics, does it just lie in the simpler summation of synaptic weight, or is it due to more-difficult-to-reproduce nonlinear synergetic effects? It would be premature to attempt an answer. All we can say is that studying the abstract qualities of space via topology, graph theory, and affine geometry, has a beneficial effect upon meditative practice: because whether or not these mathematical modes turn out to be fruitful theoretical paths for the future of neuroscience, there's something strangely familiar about every geometry which *forgets the metric*. What we can do is ground ourselves in the facts of anatomy, pay attention to physiological details, and with an ever clearer head investigate subjective experience while resisting wishful conclusions. All ambitious phenomenology has been so *soaked in the wishful* as to be useless. For example, Buddhist psychology seems at first to promise rigor in its insistent decomposition and rejection of the atomic self, but immediately devolves into *Abhidhamma* elaborations of a crowd of hypostatized entities just as illusory as the *atman* ever was. While Heidegger at a healthy distance seems to be doing nothing but reviving childishly conceived "categories of the noumenal" in the spirit of the neo-Kantian. What will save us, is a *menschlich* sense for anthropology, evolutionary history, ethology, and a sober sociology of the *allzumenschlich* such as we've already tried to initiate. What we want is to mine these questions from both sides, to auger both up and down into the mysterious core of neuromorphic experience: this is why residual processing artifacts, peripheral hallucination, errors, injury, psychedelia, trance, misconception, and illusion are so useful to us, because they hint at the invisible scaffolding. And those of us who have spent half a lifetime staring backwards into that shaded orifice guessing at its faint outlines, and learning to control our hallucinations of the imputed structure, are better prepared to conjecture about this *dark kernel* than anyone else. This is what music speaks to when it moves us – not merely superficial handles of the limbic system, not merely sophomoric sentimentality, but resonances in the deep processing routines which constitute the vast unconscious domain.

WHAT IS MEDITATION?

\- 39 -

¬ Meditation is not consolation: if it feels good, that's a side-effect of improved energetic efficiency.

¬ Meditation is not worship: if it feels humbling, that's an artifact of the necessary demolition of conscious arrogance.

¬ Meditation is not penance: if it feels bad, that's a consequence of the accumulated resistance and the storehouse of pain which your body has become.

¬ Meditation is not morally conditioned: if it feels right, that's a signal from your vestigial instinctual guidance that you're on the right track.

¬ Meditation is not divine: if it feels elevating, that's because every gain in neural coherence will amplify the degree to which you feel and *are* alive.

\- 40 -

Good meditation will leave you feeling cleaned out, refreshed, and well-arranged. It will teach you the degree to which you know nothing, and yet are prepared to know and do everything possible: this is what it feels like to be a competent animal in its proper environment, bold and true like a bird on its first flight. Quiet, attentive, and poised.

\- 41 -

What is meditation?

It begins by stilling the body. Most practitioners sit crosslegged, but any posture is possible. Sensory response should be averaged, monitored, and gently restrained, thereby reducing the composite sensory field to a stable minimum. With this achieved the focus of attention can be moved to various places within the "semiosphere" of body, while refraining from exertion. This means the nervous system idles, and avoids forming tensions, itches, or impulses. This stillness begins to permeate all systems and provoke latent tensions to the surface.

By virtue of thalamic feedback and the surjective function of conscious routines, the mind has its own set of internalized organs – it "hears," it "sees", it "feels". The narrative conscious mind is radically conditioned by *feedback*, and hence is subject to excitation paths that may continually seek homeostasis without achieving it: we talk to a self that talks back, we feel our feelings.

Yet despite the fact that the talkativeness of the mind is the primary obstacle in the practice of meditation, a theoretical grasp of this obstacle is not only unnecessary, but the wish for mastery through intellectual effort is itself a symptom of the typical obstruction.

A feeling of comprehension is not actually conducive for good meditation, but *curiosity* and *motivation* are. Therefore a delicate balance is required: we must lure out our best intellectual qualities, even while we minimize the desire for certainty in the midst of novel experience. Demonstrating this balance is the ambition of this book.

Focus on the breath is the most effective means of ending the stalemate of incessant thinking. The mind cannot be made to shutup, it can only be witnessed, influenced by this witnessing, and subtly shifted into novel gratification schemes conditioned by shifts in the *unconscious neural topology*.

- 42 -

If attempted prematurely, an authoritative approach will also fail: as with children, if there lacks a sufficient awe, the command will misfire and make it worse. Eventually it's possible to take yourself firmly in hand, like one trains a puppy, but most modern practitioners lack this kind of self-respect. The "I" must *want* to be quiet.

- 43 -

Our world is obsessed and driven by the delivery of packets of information. But in biological terms, what is information? Nothing but *potential intelligence*: a quantum of information remains unrealized until until it is exploited in the crafting of successful behavior in a given niche. Until then it is merely traffic, merely unrealized molecular currency, merely congestion.

What we learn to do in meditation, is to *improve information efficiency* through the power of nonlinear feedback: otherwise known as "consciousness", "mindfulness", and so on. The body is a pulsating electrochemical network topology, an impossibly intricate tapestry of a hundred million energy exchanges in a couple hundred energetic currencies powering an unknown number of subroutines in an unknown number of dependent networks. Do not be deceived by 21st century pretenses to omniscience: the *glymphatic* system, arguably an extremely important central nervous system network, was only discovered this century – any honest appraisal of our situation must assume there are other significant planes of causality of which we know not. Any other attitude is decidedly *unscientific*.

But our intuition shall serve: we shall *make* it serve us. The serious students of meditation have been reporting similar experiences for centuries now: the time has come to take all of it in hand as evidence, and begin from humility like a good naturalist.

- 44 -

Good meditation is about *alignment*, efficiency gains, and coherent orthogonality between simultaneous systems. The composite nervous system can be compared to a musical instrument. Like taut strings, there's sufficient internal tension to produce a sensitivity to external impulse, which is filtered according to endogenous shape to produce a natural harmonic response. What then is meditation? It's the tuning of this instrument, and sometimes afterward a bit of careful noodling. This noodling upon a rare instrument is called 氣功, *Qi Gong*: "breath work". It exploits the uncanny side-effects that arise in meditation to produce novel configurations of sensation and feedback. The exercises emphasize both strict alignment and maximum relaxation, favoring the least extension and maximal delivery of force and stability. This leaves as much internal strength left over as possible, allowing long, controlled and fluid movements into and out of all postures, leading to a sense of abundant strength. This initial illusion quickly becomes a reality, since many small, normally unnoticed muscles become isolated and strengthened. One learns how to be maximally strong through the alignment of bone, tendon, and muscle along lines of resistance and weakness in the target, alongside a full exploitation of the physics of leverage.

The core of the body, where the majority of the strength resides, is made to do as much work as possible. The spine is given length and curvature. The pelvis is held low and active. The legs are wide, the shoulders firmly in place, the hands strong. The head is upright and alert. It is the stance in which our biped frame best functions.

In this stance Qi Gong begins, and meditation takes on a new playful tone, which alleviates the pressure of pure concentration. The unfortunate truth is that meditation as practiced by the "serious" students in Zen, and to a lesser extent in the Tibetan tradition, tends to lose touch with all sense of fun. Their practice becomes dreary. All meditative discipline involves long periods of drought, but there's too much shame present in most *dharma* and this leaves students feeling they have no right to play with their practice freely.

Meditation and its gifts are our birthright as wicked, brilliant apes. I seek to teach this art in its most shameless, innocent form.

\- 45 -

WHAT IS BREATH?

載營魄抱一能無離 專氣致柔能嬰兒

Can you keep your lunar soul from wandering? Can you make your breath soft like the infant?

Laozi, 道德經, §10

\- 46 -

Look at the problem from the perspective of the vertebrate body: what is breath? It is an easily quantifiable and controllable measure of immediate metabolic rate: thus it serves not only as an energetic valve to be controlled, and not only a metric to be monitored, but as a feedback leverage enabling other systems to influence the total activation level and thus induce *mode switching*.

For example, what is the meaning of the *sigh*? Sighing appears to be common to all of mammalia because of our characteristically fast metabolism: it serves as a demarcation zone between activation levels, inducing a stepdown operation from a higher metabolic rate to a lower. By taking in a large volume of air, the previously dominant activation level is sated; and with the sudden exhalation, the parasympathetic system is activated – an affect of rest and contentment is induced and the animal tends to go to sleep.

Therefore, that breath happens to be under voluntary control is an artifact of this exploitation as *feedback channel*: we can hold or increase our breath because mammals require fast metabolic switching, such that the behavior has lost some of its instinctual imperviability and become partially voluntary. But what does "voluntary" mean? That the *initiation threshold* for the behavior is sufficiently low, such that low priority routines like apperceptive consciousness may influence its functioning. This is unlike the heartbeat, for example, which may be influenced but only through an indirect channel requiring the total lowering of metabolism: in other words, a very high activation threshold which "consciousness" alone cannot overcome.

Is it understood why I avoid the usual idiotic language here? The "ego", the subject, the consciousness, does not emit voluntary agency like a beam of light out of a magic rectum: all bodily phenomena is emergent phenomena and must be treated as such. You will not master the breath nor gain its power over your nervous system as long as you persist in such stupidities: the breath must teach you, moment by moment, the degree to which "You" are *not* in control, along with the astounding degree to which the body *can be controlled*. Resolving this seeming contradiction lies right at the heart of the practice of meditation.

- 47 -

Serious meditation practice will teach you the degree to which "free will" is an illusion: *thoughts think you*. But it is nonetheless an important illusion and not as culturally contingent as it may seem: the human creature needs the illusion of conscious agency, because it needs the *superficial coherence of narration*. Human beings are adapted to adapt as hunters of *any* ecology on the planet: the cost of this flexibility is instinctual incoherence, technological dependence, and the risk of longterm maladaptation.

The *voluntas* which moves the breath, has nothing whatever to do with this narrative noise: the several forms of consciousness are more often antagonists than cooperative. Much of the art of meditative practice is to get the narrative voice to settle into a terse compromise with apperception: this is what's known as "earnest" practice, inspiration, "meditative stability". In other words, the short-term gratification paths with which narration is generally concerned have been partially merged with the deepening trails of apperceptive routines: you are learning to find apperception more gratifying than narration, by allowing *narrated apperception* to generate longterm reward schemas. There is no shutting down of the narrative voice because it's motivated by the core human adaptive package: you must instead shift your centers of gratification and learn new behavioral routines for which the narrative function can work. Eventually narrative consciousness can enhance the apperceptive function, once the task and its rewards have been made clear: your narrative voice is like an anxious pedantic bureaucrat with too much power to be fired – you want such an agent on your side.

This clarifies why it's so important for practitioners to learn to talk about their practice with powerful vocabulary. Not for nothing was it the axial ascetics who invented some of the most alluring concepts of human history: *psyche* and *logos*, *prana* and *nirvana*, *qi* and *dao*. But these words no longer convince: they are assumptions of another age, transliterated gods of an antiquity we no longer sense. Only science convinces: therefore our necessary appropriation of its means. It's hoped that along the way, we shall manage also to generate theoretical perspectives a little more fruitful and honest, both more fecund and less prone to priestly distortion.

- 48 -

The medulla oblongata, the pons, the pre-Bötzinger complex, the phrenic nerve, the vagus nerve, the various overlapping respiratory centers and their feedback paths – none of this is necessary knowledge, nor will it be fruitful to memorize their purported functional grouping, nor to attempt to reduce the power of meditative discipline to an anatomical linearity: that would be *cargo cult scientificality*. It would obscure what really matters, which is sitting down and practicing, getting real and finding benefit: using your whole heart, your whole mind, your whole body in one moment to become a better human creature.

For the purpose of justifying the method to your intellect, it's sufficient to know that respiratory control lies largely in your brainstem, right at the very root of both the central and peripheral nervous systems, in the neural hearth. Respiration is special among the autonomic functions, because it must never cease, must possess a rigorous autonomic rhythm, and yet be subject to so-called "voluntary" intervention. Consider the case of the canines: in order to use that wonderful nose, they have to interrupt the autonomic respiratory loop, and introduce the rapid inhalation called sniffing. For a gibbon to call out his territory in the early morning hours of a Thai rainforest, he has to take a deep breath and force exhalation through his larynx with just the right constriction. This is all "instinctual" in the ethological sense, but in neurological terms it's "voluntary" – in that it's episodic, brief, and an interruption of the autonomic process. So again humans aren't unique in the control of breath, and even our linguistic capacity is clearly built on earlier forms of vocalization: the monkey screeching in the tree is also using "breath control".

So although I'm fond of the old metaphors of प्राण *prāṇa* and 气 *qi* – the idea that breath is cosmic substance, that it is the food of the gods, that it is the seed from which all existence springs – and although there still might be some obscure truth in this, and although these are terribly beautiful and even useful ideas, in a strictly reductive sense there's nothing special about *gas exchange*, except that the neural mechanisms involved provide a powerful opportunity for neural modulation. Breath control is what makes consciousness look so impressive, the big gleaming hockey trophy in the center of the foyer, without which being as *ridiculously conscious* as humans have become, would be half as rewarding and feasible. Sometimes we talk about meditation as though it were something exotic, but neural modulation through breath is also common knowledge and employed at every moment. Take a breather.

- 49 -

I'll say it again: without respiratory neural modulation, apperceptive consciousness would be rather pathetic in our hands. We'd be very different creatures, probably much less conscious, because we couldn't afford the risk without this safety net. The power of meditation doesn't make you divine in the sense of "otherworldly". It makes you divine in the sense of natural, alive, whole. It teaches us to stop using more consciousness than we need – otherwise known as anxiety. It makes us alert, poised, awake. Perceiving but not commenting, feeling but not elaborating, vigilant but conservative. So much of what makes us weak, clumsy, laughable, is *misplaced apperception*. Thinking about thoughts, worrying about feelings, anticipating events, fumbling with phrasing, refolding fraying origamic fears, exhuming favored crumbling gripes, watching and waiting for something that will never happen and is impossible. Apperception clumped and tangled in a ball just behind the eyes: that's the hypermodern condition.

People sometimes ask me how come I'm so smart, and want to know whose books, what words, which theories. But that's not the secret. And yet y'all don't believe my secret: when I tell you to go for long walks outside, to stop trying to be so smart, to shut it down and give it a rest. Learn to exhale. What you're not seeing is that in every interval, I'm striving to be as *dumb* as possible. I just take a few more steps, a few more breaths, and wait for something interesting. I strive to use as little consciousness as possible to achieve my ends, to leverage unconscious intelligence wherever possible. To not know, to not believe I know, to not need to believe I know. To find the space of a question, to wonder at the space of a question, to be wondrously dumbstruck by the space of a question.

- 50 -

Allow me to speak briefly to the more experienced practitioners, as an example of the kind of vocabulary we should be developing:

The "kundalini shivers" are associated with *inhalation* and thus chaotic excitation patterns, while the myoclonic cascade and pseudoepileptic microseizures are associated with a higher degree of order in a local minimum, and thus coherent oscillation usually achieved via *exhalation*. They are both forms of neural discharge, but in my experience quite different in character: inhalation accelerates the system into a chaotic stuttering, exhalation tends to organize neural activity into a dominant stable waveform.

This is not as esoteric as it seems. The kundalini shivers are akin to what happens in *yawning*: a trembling associated with hyperoxygenation, and probably also therefore the "PGO waves" of REM. The shivers strike me as a kind of upper harmonic, a stuttering along the surface, something like an upper resonant frequency which does not discharge the fundamental of the excitation because it remains too chaotic at lower registers.

- 51 -

Yet despite these tantalizing details, the usual neuroscientific approach to meditation yields *less than nothing*: not only because neurology is still little more than the naïve cartography of ill-defined superstitious abstractions, but because it has nothing whatever to do with *praxis*. Existing neurological accounts of meditation are about as useful as when Descartes localized the "seat of the soul" in the pituitary gland: as long as our psychology remains wishful and immature, all the chemical and electromagnetic mastery in the world won't bring us a single step closer to understanding our own experience.

Yet I believe that a *strictly physiological* account of meditation is the most useful place to begin. Everything which can be felt and influenced should be described as accurately as possible: the diaphragmatic breath, controlled abdominal contraction, the quantity and quality of breath, the subtle influence of the heart-beat, muscular tension in the face, neck, and chest, the perceptible waves of serotonin through the gut as the parasympathetic nervous system becomes dominant, the prickling pressure in the zone of the third eye, the relaxation of the pupils, the alignment of the spine, and so on. All of these elements may be experienced, tested, and refined: *thus* is the path to "neurology" – that is, an actual mastery of one's own neurochemistry and the balance between the various interdependent systems. In other words, the *fruits* of the practice yield understanding, not the other way round: clairvoyant theory is the last element to arrive, not the first. The intuitive control over your own body which meditation will give you, far outstrips anything a merely intellectual exercise can offer. Anything else is like describing music rather than listening to it – or better yet, *playing* it.

- 52 -

Thus it is that Katsuki Sekida, in his work *Zen Training* in which he attempts a scientific approach, proceeds from the third chapter full of premature neuroanatomy, to an excellent description of the physiology of breathing in the following chapter. Because it's based on personal experience and practice, it's even worth quoting at length:

> The instruction to oppose the contraction of the abdominal muscles by the diaphragm sounds complicated. In fact it is very simple: you have only to hold your breath. If you then expire slowly, little by little, it is necessarily done by holding the diaphragm down and steadily checking the upward pushing movement of the abdominal muscles. The latter are placed in opposition to the diaphragm, and the contraction of

each is increased. This is what we mean when we speak of "throwing strength into the tanden."

In normal respiration, when the lung volume reaches 2800 milliliters, inspiration automatically turns into expiration; the inspiratory muscles relax, the volume comes down to 2300 milliliters, and all the tension returns to zero. Normally, inspiration then automatically starts to take place again. In zazen, however, you do not stop expiration at 2300 milliliters but continue to expire, and this calls for effort. In general, then, we may say that above the horizon of breathing it is inspiration that requires effort, while below this it is expiration. Normal respiration is performed above the horizon, using the tidal volume alone, and expiration comes about by the relaxation of the inspiratory muscles. In zazen, expiration goes down below the horizon, and it is in this phase that most effort is exerted. It is this expiration below the horizon that is principally effective in bringing about samadhi, because it is here that the diaphragm and abdominal muscles are brought most strongly into opposition.

Without cooperation of the respiratory muscles you cannot move any part of the body, pay close attention to anything, or, indeed, call forth any sort of mental action. We cannot repeat this fact too often: it is of the greatest importance but has been rather ignored up to now.

Sekida, *Zen Training*, ch. 4

\- 53 -

And yet we also want – and seem to need – more intellectual promise than austere physiological testimony. We need a plausible theory of the *adventure of mind* – which was honestly why we began this journey in the first place: we want a viable reductionistic theory of the subjective thrills, we want something that will not only inspire us to meditate, and not only help us become who we want to be, but a story we can be proud of in the 21st century, a story freed from metaphysical presumptions and yet imbued with the unsolved mysteries of being-alive.

This is why I've begun dropping bits of mathematically infused candy along the path: I hope to demonstrate a "postmetaphysical" relationship to subjective experience which avoids the self-flattery of clandestine recoveries of metaphysics so typical of postmodern philosophy – e.g. Deleuze, Derrida, Lacan – even while we allow ourselves *to dream* openly, unashamedly, like the desert nomads we were meant to be. We want the right to carry relentless rationality into the very heart of mysteries, like the Delphic cube sitting atop the steamy abyss, tantalizing our pride.

- 54 -

I can tell you from experience that when the Daoists talk about "stilling the breath", or even ceasing it, it's not mere hyperbole: it is possible to stop breathing in the usual sense, without losing consciousness. This is accomplished by sitting upright with the spine well aligned and the diaphragm lowered, slowing the metabolic rate to the minimum, gradually slowing the breath while maintaining a light optimal muscular activation along the chest cavity and lower back, which must be well developed, and eventually becoming so still in the body, that the ambient undulations of the diaphragm and the vibrations of the heartbeat itself, are sufficient to maintain air exchange in the fully distended lungs. Upon closer inspection we may discover that the diaphragm is in fact in motion with a constant gentle undulation, like rippling water. This is the anatomical description of deep meditation as I've experienced it.

What happens in the mind during all this? All kinds of distraction, but a stillness of the body this profound will eventually triumph – temporarily.

- 55 -

I can also report that it's possible to allow the body to fall asleep without losing consciousness. However, it's difficult to withstand the panic without disturbing the delicate process, because you will experience that "infinite falling" sensation as your voluntary motor control is shut down. This is the origin of that rather commonplace experience of jolting yourself awake from a shallow sleep. But with sufficient control gained through meditation practice, entering a dreamstate directly from waking is possible. For the especially dedicated student of meditation, this is not such an unusual experience and is in some schools one of the expected markers of progress.

- 56 -

Do these powers sound impressive and mystical? They are not. Don't imagine that any of this is unique to that precious fondled creature called the human being: any fully healthy creature is capable of all of this and more. Cats and lizards, for example, freely drift between various states of trance, hypnagogia, and tightly controlled metabolic ebbing. Only as long as we keep our attention fixed upon *the unbelievable miracle of the vertebrate nervous system*, will I admit that these powers are "mystical". My hostility to anthropocentrism is much deeper than any gratification gained from imagining these are special powers: in fact, by insisting that meditation is *remedial education* in our animal birthright, I believe I gain something much more important and more deeply gratifying – not another metaphysical buffoonery but a true homecoming.

WHAT IS BREATH?

WHY DO WE MEDITATE?

\- 57 -

Our contention is that meditation calibrates basal neuronal function: it tonifies the limbic system such that you freak out less; it calibrates the proprioceptive system such that you become less clumsy; it clarifies the boundary conditions of parasympathetic activation such that you spend less time oscillating between poorly defined rest and activity; it retrains thalamic modulation of narrative consciousness such that the reward calculations so typical of the forebrain realign to healthier vertices. All of which means we become calmer, more competent, more decisive, more energetic, and more willful and purposive in all things. I lay emphasis on the fact that meditation is effective to the degree that it targets the *lowest-level* and thus *highest-priority* systems: *the effects percolate upward*. Meditation is sanctimonious garbage to precisely the degree that it promises to make you holier-than-thou, that it targets your "highest" nature, that it encourages fantasies of transcendence and otherworldliness – in other words to the degree that it targets *social reward calculation*, unresolved parental imago and Oedipal entanglement, and everything which hints at penance and reward, trial and judgment, acknowledgment and love. Meditation is not about proving yourself worthy to a withholding god, nor besting the haunting voice of your father, nor winning back the first embrace of your mother. You will never win that game, because the solution is *not within the game*: these hangups will arise in meditative practice with a renewed force, but they must be attacked from a different angle – namely symbolic analysis. Part of our ambition is to point the way out of the snares laid for millennia by the priests: they claimed ascetic practice and made it their own; they perverted the exploration of the bleeding edge of human potential into just another means of instinctual castration. This is why the early Daoists are such a precious testimony: they were relatively untouched by the priestly agenda and maintained the pride and rigor of hermitage as an antithesis to priestly ambitions. Why solitude? Why is solitude an essential ingredient in the recipe of maximal human development? Because social reward calculation is our stupidest, loudest voice: it makes us into abject groveling fools. It preys upon our deepest wounds. Living in a state of civilized tribelessness, such as we do, exposes this fundamental weakness of human character: only a deep and abiding acceptance of isolation can defeat it. What is generally not known or expected, is that just as often, an almost unheard-of *richness of companionship* lies on the other side of such solitude: there are more of us than there may seem.

Why do we meditate? Because we want to become something more. Why do we hesitate? Because we fear the consequences. Growing up into the containment of our potentialities, fleshing out the skeletal frame of our emotional constitution, growing accustomed to the sensation of productive freefall can be just as frightening as not getting what you want – indeed much more so. Because what comes next is undecided, uncharted, unfelt. Most of us prefer the worst of what we have already known, to the possibility of a novel indeterminate experience. Therein lies my last claim concerning the powers of meditation: it will strengthen your tolerance of *ambivalent novelty*. It's possible to trace the trajectories of pleasure and pain, joy and sorrow, desire and repulsion back to where they first diverged: there is the ecstasy of being alive.

- 58 -

Neurochemically there doesn't seem to be anything quite like meditation. It's like waking from the perfect nap without grogginess. The stomach is settled, the muscles are relaxed but ready, the mind is clairvoyant and childlike, the mood is buoyant and receptive, yet also serious and mature. I can't list the ingredients of this neurochemical cocktail – but I can tell you how to cook it.

- 59 -

Emotional reparations gained from meditative practice: to provide a "holding environment" as the ego psychologists say, in absence of the kind of emotional stability one should have had in childhood. To learn to deal with the sense of perpetual loss and abandonment which characterizes modernity. For some of us with too much integrity and introspection to have been *candidates for narcissism*, we learned to develop the meditative arts as a kind miniature community – never losing sight of the fact that the primary function of so-called "narcissism", is to generate a small manageable tribe: the gazing lover, the beloved image, and the threatening reality. In vigorous meditation, you have the practitioner, the practice, and everything which resists the practice: in other words, your *devotion*, the *ideal*, and your perpetual *failure*. So much of what's valuable in genuine meditation instruction, are the *tales of successful failure*: or how get the ideal to switch places with your failure, and the failure to switch places with the practitioner. Good meditation happens when there's a kind of "tensegrity" between all three elements.

- 60 -

I have meditated seriously and daily for more than 20 years. The first 2 years of practice changed my life permanently. I set up a little hermitage in my overeager youth, and aside from earning a little money to support myself – as a construction worker comically enough – I practiced nothing else. On good days I would manage an honest 3 hours of concentrated practice. I followed a cycle I could repeat as often as possible: exercise, bathe, meditate, eat. For a brooding, overintellectual, uptight white boy – it was a revelation. I discovered spontaneity and physicality, I discovered an emotional life, I discovered self-love, I discovered the intoxicating power of concentration and stillness. But eventually it faded, and like all intoxicants the highs no longer satisfied. I grew intensely restless: doubt and disenchantment began their long campaign...

- 61 -

Years later, I still practice. Why? Because it's *a form of hygiene*. Meditation for the all-too-modern human being is a matter of spiritual cleanliness. Nothing is so necessary to the overwrought mind and the underutilized body as a little vigorous stillness. Thus I insist that the benefits of meditative practice have more to do with *refinement of unconscious process*, rather than yet more overexploitation of conscious acrobatics.

Our best spiritual development will not be found within yet more anxiety-conditioned self-awareness, yet more half-therapized consolation, yet more moral posturing before a mirror of blamelessness. We must learn to properly align and exercise the system *as a whole*: what we want is the analogue of "athletic heart syndrome" in our bodymind. How do you obtain this kind of easy stillness? Not by "abandonment of desire", nor "insight", nor any moral obedience – but by exploiting nonlinear emergence via conditioning of the many unconscious subsystems, which produces a far greater yield of intelligent behaviors at the conscious niveaux than the usual topdown approach.

- 62 -

Insight is not the *cause* of enlightenment, it is a *product* of health.

- 63 -

Meditation is no panacea. But according to even the most trenchantly hostile cognitive psychologists, there are a few uncontested benefits of meditative practice:

- Reduced stress
- Improved anxiety tolerance
- Enhanced learning ability

But if that were all it accomplished, it would not be worth its mystique – and we might as well substitute pet ownership and a daily exercise routine: which is precisely something agents of the status quo would like to insinuate. But they'd rather see you a *drug addict*: have no doubt that their pails full of pills will be brutally effective...

- 64 -

Why do we meditate?

¬ It stills the blabbering narrative mind: overcoming the tendency to fill every ambiguous space with useless talk is the first challenge.

¬ It awakens the body: an enhanced proprioceptive capacity makes "being-in-the-world" more interesting than a droning monologue.

¬ It repairs an overloaded nervous system: unlearning the habits of repressive psychosomatic quarantine, replacing afferent garbage-hoarding with informative parsing and efferent garbage-disposal, is one of the crown jewels of meditative practice.

¬ It teaches intuition: a cultivated attitude of curiosity, patience, and humility before the present moment makes us much more receptive to *unconscious intelligence*, and thus the totality of latent neuronal brilliance, than the typical arrogance of the loud, clinging, noun-munching conscious posture.

¬ It develops emotional independence and equanimity: the more often the chasms of indeterminate multivalence are faced, crossed, plumbed, the less we panic in the presence of novelty.

What are its contraindications?

- It tends to induce hypervigilance. It can initially worsen the overactive mind and under certain circumstances, lead to a greater proclivity to anxiety: the collapsing roof of an overly ambitious structure with bad foundations – a commonality among some of the most talented students subjected to the bad-faith teachings of the priests, who need to enfeeble their strongest rivals.

- It can lead to obsessive self-monitoring. It encourages the illusion of control through maximum sustained consciousness: in the mediocre students, this leads to premature delusions of mastery.

- Genuinely exotic experiences can be abused to encourage fables of "contact with divinity", absorption into "pure consciousness", and other fantasies of transcendence and spiritual hierarchy.

- 65 -

What does meditation impart? A growing sense of agency, a deepening appreciation of the body, a burgeoning gratitude toward the shimmering world, an unexpected mastery of sensorium and mood, and a rare relief from the sense of *victimhood* which has a contagious quality in our age. Let me say it again for the hard of hearing: sitting still and paying attention to your breath, will disincline you to *passive lazy behavior*. It teaches *discipline* at a level to be found nowhere else, because it demands it at a level to be found nowhere else: thus am I so hostile to the portrayal of meditative practice as just another *consolation* of a helpless wretch, just another stopgap in a hopeless case, just another day at the spa, just another indulgence of a bored brat.

Meditation is often misunderstood as a kind of stupor. But it's a willful mischaracterization: there's a long history of Eurocentric slander of "oriental laziness" aimed at this discipline, which would like to hide its own terrified workaholic escapism behind the name of virtue, and employ as many gravediggers as there are ambiguous feelings, desperately avoiding contact with stillness because it is so ambivalent and potently associative.

True sitting meditation can be the most difficult task of all: actively, willfully sitting still and paying attention to breath and body. Even 20 minutes of this discipline, daily and successfully performed, requires years of practice. Dedicated practitioners know only too well that years can slip by "on the cushion" without any signs of progress.

- 66 -

Moreover, *meditation is easy to fake*. Most of the *braggarts* in this art – and they are the rule – get no further than endless discursive thinking, rambling and reactive feelings, gratifying fantasy chasing after fearful anxiety, a chaotic body alignment, and so on. Realistically assessed, a 30 minute session is often highlighted by a two minute stretch where one finally settled down and decided it was time to meditate. *One minute* of total concentration is enough to give the beginner extraordinary visions and ineffable feelings. If you don't believe me, try it. Be honest with yourself and observe how long you can remain focused without beginning an internal dialogue. It will be a *very* short time.

- 67 -

Where are these "siddhis", these magic powers we were supposed to obtain? But relative to the average modern slob, we *do* obtain magic powers: freedom from pointless worry, the ability to know what one is feeling at any moment, the presence of mind and body not to make stupid mistakes, the quiet of mind not to commit unconscious acts of neurotic revenge on oneself and one's family, the ability to know what one wants, the courage to allow annihilating insight, sufficient experience with spiritual vertigo and the sense of falling apart such that one no longer resists it, and of course the ability to read other people like the open books they really are...

But even after all that's in place, there's still the much greater difficulty of giving oneself *permission for this power*: what will we do with it? To what end? Everything depends on trusting ourselves to be responsible and worthy of the imbalance of power, whether we feel "compassionate" or not...

- 68 -

What is meditation? Remedial training in physiological homeostasis. Cognitive expansion through entrainment to the rhythms of the breath, the heartbeat, the musculature, and the digestive tract. To learn a skillful balance and responsiveness to the many sources of information of the body-collective... All of which we were supposed to have learned as children, and all of which any healthy animal in the wild knows to its bones. To work and play hard, without ever straining oneself; to learn how to rest frequently and at every available moment; to learn how to minimize effort wherever possible and only maximize at the critical juncture... This begins to sound like training in martial arts, and perhaps that's precisely correct.

- 69 -

> **常にも兵法の時にも少もかはらずして心を広く直にし、きつくひっぱらず少もたるまず、心のかたよらぬやう心を直中に置て心を静にゆるがせて、其ゆるぎの刹那もゆるぎやまぬやうに能々吟味すべし**
>
> In both everyday and military events, your mind should not change in the least, but should be broad and straightforward, neither drawn too tight nor allowed to slacken even a little. Keep the mind in the exact center, not allowing it to become sidetracked; let it sway peacefully, not allowing it to stop doing so for

> even a moment. You should investigate these things thoroughly.
>
> Miyamoto Musashi, 五輪書, 水之巻

Meditation is remedial training in feedback modulation. It reinforces our most ancient instincts of *homeostasis*: overlapping intertwining rhythms of breath, heartbeat, peristalsis, the subtle undulation of mood, the "gently swaying mind".

Meditation restores our birthright to the healthy, well-rested, well-exercised body which every exuberant puppy, every napping cat, every flitting sparrow, every lizard in the sun practices every day. While it's true that we thought we were climbing a mountain, when it turned out we were only digging our way out of a hole – that does not mean however, that this plateau of health should be considered "normal" in any modern sense! I will forever advocate for the "enlightenment of animals" as the proper paradigmatic orientation of our spiritual life. A quiet but alert mind, a relaxed musculature, the ability to fall asleep whenever necessary, the freedom from excessive inhibition and the readiness to fight, flee, mate, care, abandon, eat, and rest – all in proper proportion.

What we get in meditation is a foundation of subjectivity more stable than the echo of the last conscious sentence: a glowing ruminant interoception, the topological undulations of proprioception. It comes *not* from without or "above" or "beyond" and it definitely is *not* "consciousness itself", it is the peace of the healthy intestines, the functioning liver, the muscular heart, the confident idling nervous system... Only in *this* condition and no other, can the as-yet-unknown potentialities of maximal human development be explored: only via a renewed homeostatic foundation, can we find out just how smart we are and what we want to do with it.

THE GENEALOGY OF MEDITATION

- 70 -

I hunt with dogs. Often in the midst of a long walk through my desert landscape, I will stop to meditate. My dogs can discern the quality of my meditation: if it's a deep trance, they sit quietly nearby and maintain strict vigilance. If my mind wanders and chatters, they grow impatient and tell me to move on. The ability to meditate such as we know it, derives from the predatory ability to *stalk* – which is itself probably derived from the prey's ability to hide. The hominid has only achieved a partial transformation into predator: we are technological cheats who snuck past the instinctual refinement all other predators had to undergo. Thus our meditative skill is tenuous, fraught with error, prone to degradation into anxiety, vanity, and fantasy. We are half mad, the ape who talks to itself, who sees things that are not there: sometimes we are genius, more often we are barely sane. Meditative practice is the restoration of the predator's sanity, the exploitation of predatory vigilance for the purpose of taming our liability to anxious neurotic nonsense.

Our ancestors once gained this sanity by actually hunting, gathering, and crafting: we "liberated" moderns have to simulate these necessities with meditative discipline. Most of us, however, do not possess sufficient *inner necessity* to achieve anything significant: most meditation practice you see and hear of, is *falsified*. Meditation is easy to fake. It begins by lying to yourself about the quality of your practice, but once the rewards of priestly authority appear, one quickly learns to lie to everyone else. It's best not to trust a teacher who speaks of meditation in terms other than *constant failure*.

- 71 -

I have found that stalking prey makes for excellent meditative practice. The emotional tension and the concrete challenge eliminate many of the chronic problems associated with sitting meditation: a wandering, chatty mind; a sense of pointlessness; restless boredom. As long as the possibility of a real confrontation with a wild animal persists, the human body will naturally remain poised, alert, and serious. The mind will be occupied and obedient.

The animal need not be any larger than a sparrow, and one need not actually harm it. But the stalking intent must be real. I first discovered this practice many years ago, while doing Qigong: I found that squirrels could not see me. And often when they did perceive a human presence, the practice rendered me very curious to their eyes: they became inquisitive and would often cautiously approach, attempting to ascertain what this unknown animated thing could be.

Over many years I eventually put it together: the human capacity to hunt, is predicated upon *the primate talent for mimicry and subterfuge*. We are supreme actors, impressionists, liars, or if you prefer, "shamans". We have the ability to modulate our spirit, which properly speaking means the animating forces of our body, which reductionistically speaking might mean the maximum attenuation of amygdalic and sympathetic nervous system activity despite the possibility of danger. We are capable of hiding our intention in plain sight: we are masters of the mask.

This accounts for why it was that I often had so many encounters with wild animals in places others did not: more than once, I have encountered bears in forests others considered empty of them. But this is because even as a child, I had the habit of walking quietly alone, lost in my thoughts...

- 72 -

Once, after a productive meditation session in a rich patch of Bavarian woods, I was walking home and had paused in the middle of a path, admiring the forest view. Striding down the path came a large grey fox, as confident and at ease as you could wish. I didn't move until he was nearly upon me – maybe six feet away – and I became a little concerned that we might actually collide. He did a double take, looking very unfoxlike in his surprise, and darted off. Strange! For a creature as circumspect as a fox, not to see a human being standing in the middle of a path with no camouflage whatever may sound terribly unlikely – but the story is true and it's not the only one.

- 73 -

One of the odd practices I have learned in my adopted lifestyle, is the trick wherein one moves slower than the threshold of the prey's perception. For most mammals this seems to reside at around one inch per minute. The way to move this slowly is to use the rhythm of breath: with every inhalation, allow a very slight muscle contraction. What will make you particularly invisible to an animal, is the ability to move like a plant: to sway gently, and move your limbs stiffly. The principle difficulty in all this, is *hiding your intention*: you must keep your predatory gaze subdued, only a faint glimmer. Most prey animals will be suspicious of the situation not only because they will likely catch whiffs of your scent even upwind, but because of this atmosphere of a glimmering murderous intent: but most prey must deal with a constant skittish paranoia anyway, so if you are patient and still, they will quickly dismiss their own intuition and go about their business.

- 74 -

I am by no means so unusual in this capacity to hide myself in plain sight. I once knew a girl who could make herself socially invisible: very useful for getting into parties without paying the cover charge. I saw a bouncer look directly at her, and yet he seemed not to see anything but a mirror: she was able to turn herself into a kind of placid reflection, present and yet nothing there but one's own noise.

Highly intelligent women are social chameleons. The female ability to practice meditation such as I know it, seems partly inherited from the male hunting practice, but perhaps more so derived from the necessities of tribal life and tribal warfare. There is a reason why young women were always considered prime material for oracular divination.

- 75 -

There is a story about Marilyn Monroe: that she had the ability to turn her "thing" on and off. She was out with a friend in the city, and her friend remarked that no one seemed to be noticing who she was. Marilyn explained it was because she didn't want them to, and then demonstrated: without changing anything externally, she turned it on. Suddenly bystanders recognized her.

- 76 -

The evolutionary history of meditation:

1. The rodent talent for nocturnal theft. The rat.

2. The newly acquired ability of a hunting primate to stalk prey. The chimpanzee.

3. The Paleolithic hunting technique of disguising oneself as an animal among animals. Pashupati.

4. Trance as amplifier of the ritual context, the exploitation of weakened instinct for the production of exotic states of mind as marks of shamanic power, magical efficacy in tribal psychological warfare, the blessing and the curse. The shaman.

5. Linguistic suspension, and the effect it has on a creature ever more dependent on linguistic mediation: "suddenly I understand". The sage.

6. A remedial education in peak bodily performance in a postagricultural world: waking up the somnambulistic civilized body. The Daoist.

7. Cultivating an aura of holiness in order to overawe the common people, so that they will feed you. The marks of self-harm as contributing to the aura of otherworldliness. Sadistic cruelty as the mark of the holy man: the one who has sacrificed himself to "pure consciousness". The sadhu.

8. An education in repression and its exploitation by the priests, as another social program of instinctual inhibition. The Buddhist.

9. The synthesis and recovery of shamanic power in late modernity, meditation as therapy, meditation as neuromodulation, meditation as birthright: our 21st century practice. The apocalyptic hipster.

- 77 -

It will surprise some readers to see meditation given such a strange and humble origin story: the sneaky protorodent, vigilant when dinosaurs slumber; the wary grassland chimpanzee, stalking his antelope; the ice age hunter, flopping on the ice imitating a seal while creeping closer to the pups... But I believe meditative discipline needs no flattery, and will benefit from the humility to see its genealogy laid bare. Human greatness has its origin in *weakness*: that's our riddle, our burden as the techno-ape, which we will not learn to overcome until we embrace our absurdity and disappointment. I've driven away more than one student already with this story, but in my view the emotional capacity to *withstand disappointment* is not just the entrance fee but part of the reward itself: only in the aftermath, does the world begin shining again, reborn as a ridiculous, unexpected, resplendent comedy. Meditative discipline as the growing appreciation of the *comedy of errors* we are born to – not only civilization but the evolution of life itself as a comedy of errors, as a rambling rumbling tangent slowly becoming something grand and eloquent precisely because of its errancy, precisely because it's the fool who stumbles into wisdom.

PART II:
SENTIENCE

THE NEURON WAS BORN TO SWIM

\- 78 -

The assumptions of Von Neumann architecture infect contemporary neuroscience: it's implied and assumed that the reasoning mind receives *mere data* from the hindbrain, as though everything below the cortex were just buses, routers, or at most sensory transducers. What we must internalize is that *there is no central processor* in any organism: organic structure is *radically parallel*, redundant, robust. Even the nucleus of a cell contains many various pieces of the genome under many various stages of transcription: every "singularity and manifestation" model of life is a bad metaphor, which has had too much currency for too long. Life breathes synchronicity into the asynchronous, life is convergence, life is emergence: when will we be prepared to accept this axiom in the field of neuropsychology? We must stop looking for imaginary entities dwelling in imaginary shrines: there is no ego, no unified personality, no self, no hall of perception, no unity of consciousness, no homogeneous episodic memory, no static identity. Every seeming unity of purpose and will is an emergent order, and generally the more focused the behavior the more energetic the dissipative flow: a creature that seems to know what it's doing is a creature *compelled*, a creature *tyrannized* by its own vitality.

\- 79 -

> Biological systems run on massively parallel, low speed computation, within an essentially fixed topology network with bounded depth.
>
> Rodney Brooks, *Intelligence Without Reason*, §3

The calculations performed by the cortex do not operate upon static datasets: they operate with *behavioral outcomes*. The purpose of "mind" is not to determine the objective "Sachverhalt" to lisp like the young Wittgenstein, but to determine *what to do*. This determination does not need objective modeling, and judging by the history of philosophy it seems to grossly err with every such attempt: it only need weigh the relative weights of its contingent behavioral thresholds against an accurate but extremely specific sensory array. A living creature is not a reality reproduction machine: *intelligence has nothing to do with objectivity*. A living creature *inherits* successful behaviors from its ancestors, and its job therefore is to balance them delicately and wisely, but perform them with abandon: much of what it means to live is to do justice to the badass attitudes of our forbearers – whether you are a Pacific salmon jumping up a cascade, a chipmunk stealing dogfood in broad daylight, a shrike impaling a grasshopper midflight, a snowgoose migrating in a thunderstorm, a dormant scorpion waiting in eternity for spring.

It's not that scientists have been projecting their own *modus operandi* upon the subject of intelligent behavior: they've been telling the white lies of sociality. "Collect all available data, reason and plan in abstract semantics, and translate into action": this is the fable of free will, this is the fable of conscious agency, this is the fable of personal responsibility and thus guilt and thus punishability. When it succeeds, the ape says: "I meant it." When it fails, it says: "I didn't want it." When it perpetrates, it says: "I didn't do it, and anyway I had a right to."

- 80 -

There is no distinction between data and computation: biological computation is a kind of side-effect of behavioral loops. There are no truly abstract semantics: there is no "pure thought", only elaborations of premotor and perceptual processing. That we believe in such a ridiculous thing, is an artifact of the influence of postaxial philosophy concerning the primary role of *narrative consciousness*: we want to believe that the auditory hallucinations we all know and love and loathe, are the last veil before the holiest of holies – the seat of "thought itself".

- 81 -

> We are interested in when a neuron or neural system evokes an action or makes a decision. [...] If there is no choice, there is no effective information.
>
> J. Feldman, "Ecological expected utility and the mythical neural code"

The purpose of mind is not "objective modeling", it is to produce successful behavior. It accomplishes this not by reasoning about the world, but by virtue of a balanced and calibrated structure – which is not essentially about modeling the state of the world, but producing coherent successful behavior in response to stimuli. The *Merkwelt* of any given creature is likely to be highly distorted, because it is always highly *valenced*: all information is weighted by virtue of the creature's structure, not its "reasoning". Thus a great deal of what we're doing in meditation and spiritual development, is adjusting the weights of a perceptual field. Slowly tuning down amygdalic covalence, slowly leveling a plane which was once highly graded, such that many paths become possible and we begin to become a little unpredictable. Spontaneity is ours again: which means we get to become funny, obscene, and eloquent by turns. We should be like a spinning top: lightfooted, elliptical, and difficult to displace. An angular momentum that keeps us balanced, a lavish expenditure that makes us energetic: part of my definition of a spiritual life is that it seems to obtain energy from nowhere. This is some of what has made all the talk about "higher planes" and transcendent effulgent realms from which we feed, seem a little annoyingly relevant to those of us who know: but we can now recognize these "higher" things as nonlinear dynamics, efficiency gains, and the very heart of organicity itself. It's not uniquely human, and I'll only accept the adjective "divine" if it's acknowledged that what *lives* is divine.

\- 82 -

The corollary to the idea that there is no difference between data and computation in biological systems, is that *there is no neural code*: the meaning of any given quantum of neural information is not realized until it reaches expression in behavior – precisely analogous to the way words are meaningless outside of bodily experience. An ungrounded "pure symbol" is *purely meaningless.* To watch a mathematical mind realize the necessity of this premise over the course of a lifetime, read Wittgenstein. About 90% of all the chatter in Anglo-American analytic philosophy, 20th century linguistics, cognitive psychology and its cherished "mind-body problem", and the sexier dithering of French post-structuralism can be wiped away with this robust postulate: *meaning is derived from action*, not representation. Which is also why every attempt at a purely objective philosophy begins and ends with unexamined ulterior motive: the way to mean precisely what you say, is to treat language like a physiological exercise, and thus another domain of ethics. Every utterance should be a reflection of who you want to be and where you want to live: the original meaning of "ethos" is *habitat.*

There is no neural code: it would be unimaginably inefficient to encode and decode every signal between neurons, or between neural networks, or even in a sense between an afferent signal and its efferent muscular and hormonal realization. The neuron began as a muscle actuator, and therefore continues to speak *the language of muscles*: consider the rippling jellyfish to appreciate this – or imagine the undulation of the spine of a fish, with cross-lateral activation of limbs, the rational coherence of rhythm. The neuron was first adapted to *swim* – and I believe it still bears that mark, just as we still carry a simulation of oceanic fluid inside our terrestrial bodies. You're a talking bag of brine. – Have you noticed how the rhythms of swimming clear the head? But everything rhythmic calibrates the body, from making a coffee to walking to the corner store to shaking your moneymaker... One of the most powerful instruments I've found for musical trance induction, is the Persian *dotar*, because the rhythmic stroke is performed with minimal neuronal expenditure: one crooks the wrist inward and oscillates the whole mass of the hand, precisely as the mentally handicapped do for soothing self-stimulation – it's an extremely old primate gesture. All neuronal experience is conditioned by rhythm: yet it's not so much an overwhelming rhythmic conformance we're looking for – as though a pounding march represented an optimum – but the subtleties invoked by successful polyrhythm, even if only hinted by syncopation. Why does this matter for the student of meditation? Because when you learn to experience your inner life as a series of concentric pulsations, and the underbelly of the impulse to think as a throbbing cadence, you will come much closer not to *suppressing* it as the Buddhists want you to, nor to *indulging and abusing* it as most of us do, but to riding its waves skillfully, seeking neither quiescence nor finality.

- 83 -

Addendum. The problem is that neuroscience still looks to the seat of the soul for the meaning of any given emotion, deed, or thought. Thus it is that they're always looking to the isolated brain: they're sure that somewhere in there lies agency, choice, freewill, personality, and the ultimate referent of all interpretation. But the meaning of any given behavior arises from the totality of the organism in the environment, including the lifespan of the individual, the history of the species within its niche, the local interactions with all other species, elemental cycles, seasons, and molecular resources – the entire phylogeny of the species reaching back to the beginning. *The whole conditions the parts*, and perhaps the future as well as the past: there is no last touchstone of interpretation. Every theory, every language, every age, every personage, every attitude, every mask, every history, is *incomplete*. What's the point therefore? Of this theory, this striving, this discipline? We don't get to know – and it's only our job to fulfill our role with as much boldness, austerity, or filigree as suits it.

THE NEURON WAS BORN TO SWIM

THE MORAVEC PARADOX IS NOT PARADOXICAL

- 84 -

> Encoded in the large, highly evolved sensory and motor portions of the human brain is a billion years of experience about the nature of the world and how to survive in it. The deliberate process we call reasoning is, I believe, the thinnest veneer of human thought, effective only because it is supported by this much older and much more powerful, though usually unconscious, sensorimotor knowledge. We are all prodigious olympians in perceptual and motor areas, so good that we make the difficult look easy. Abstract thought, though, is a new trick, perhaps less than 100 thousand years old. We have not yet mastered it. It is not all that intrinsically difficult; it just seems so when we do it.
>
> Hans Moravec, *Mind Children*

The "Moravec Paradox" is not a paradox at all: it is the result of *bad philosophizing*. In other words, unexamined assumption, unconscious metaphor, and self-congratulatory conclusions have plagued this field just as they have many other fields which depend upon shallow psychologizing and naïve introspection.

Intelligence does not reach its apex and most essential expression in abstract reasoning: a *truth table* is not the heart of intelligent behavior. Proprioception, creative manipulation, and navigation in a complexly textured real world are the essential challenges of intelligence and therefore the foundation of all other forms of intelligence, including symbolic reasoning.

But how could it be, that symbolic capacity originates in perceptual and sensorimotor tasks? How could *being-in-the-world* result in *reasoning-about-the-world*? What's the probable evolutionary genealogy?

- 85 -

Let's consult ethology for a moment and consider the example of the "supernormal sign stimulus". As Nils Tinbergen famously demonstrated, an animal is not only fooled by a substitute sign, but will respond proportionally to a sign's *exaggerated* character: the nearly-blind fledgling seagull will respond all the more vigorously to the minimal but exaggerated sign stimulus of an unnaturally bright red dot on the beak of its feeding parent.

What can this tell us? That the activating gestalt for an instinctual behavior can be isolated and toyed with, because instinctual discharge is *already inherently symbolic*. As the immensely influential but still largely unknown Jakob von Uexküll convincingly showed, every animal lives within its own *Merkwelt*: there is no "raw" data, but a curved field of vectors and weighted signs.

> The animal's own world is not only dependent on what its sense organs can or cannot receive. Its sensory world is still more restricted; it is composed of sign stimuli [...]
>
> N. Tinbergen, *Study of Instinct*, p.37

Hence "biosemiotics". But other than Uexküll, I've seen no evidence in that anyone in this field approaches these questions with that valuable combination of discipline and imagination. The "sensorium" is neither a *tabula rasa* as per the vestigial Cartesian-Lockean-Kantian psychology which still dominates the cognitive sciences, nor solely composed of "always-already" signs as per the fatalist French poststructuralism which dominates the academic humanities, but the undulating inner surface of *a sphere of potential signs*. The world of the most-living is itself squirming with life: not "always-already" but "maybe-now". Anyone remotely familiar with peak cortical performance – or just serotonin overdose as induced by psilocybin – should know what I'm talking about: the inner surface of the *Merkwelt* is not merely "data-driven", it is the seething primordial ooze of meaning. Generating and sustaining emergent coherence from this sensory soup, such that a bottomup optimization escapes the *obtuse blindness* which every topdown approach succumbs to, is what should be properly considered the "hard problems" of intelligence.

- 86 -

But we should not assume that such a description of the organic fails the test of rigor: did you know that proteins *vibrate* according to their macrostructure? The processes which animate the cytoplasm should not be thought of as though chemistry proceeded like a Turing tape: it's more like a Gamelan orchestra, a throbbing cadence, a vast unthinkable symphony of ratio. After all, the *ratios* which characterize chemical reactions are due to the basins of attraction which establish the orbital nodes of electron clouds, and thus the same interval structure – otherwise known as the "harmonic" – percolates upward into the great cacophony of the organic, producing yet more emergent structure built from constraint and dissipative flow. Such is how I experience the senses: neuromorphic experience is not organized by intelligence, it *generates* intelligence from its own dalliances with chaos.

- 87 -

Sense data is a heavily weighted graph, with strong attractors and repulsors: to "sense" is to traverse a field of virtual signs which actuate in degrees. Every creature is both constantly hallucinating and overreacting, and constantly ignoring and remaining passive: the question is, what matters to *this* creature, at *this* time, in *this* place.

To build anything like a viable hypothesis of intelligent behavior requires that we internalize this lesson and allow the symbolic reasoning to percolate from the *lowest possible levels*: the more a symbol is defined in terms of *recent* and *accurate* sensorimotor feedback, the more meaningful it is. In other words, abstraction is obtained with a proportional sacrifice of relevance: most of the *apparently symbolic* behavior of a living creature is only visible due to frustration and displacement. What is *more deeply symbolic* looks like "purpose", because it's hidden behind success. When a bird builds a nest by gathering twigs, we say it does this because it's nesting time. When it tosses twigs around in a courtship display, we say it's doing something symbolic and imply uselessness: but the real symbolic moment is rooted much earlier and deeper within the avian adaptive package...

- 88 -

It may aid us to realize that most of human culture consists of the trade in "supernormal sign stimuli". *The pornography of instinct*: hunting, killing, eating, mating, nesting, brooding, and so on are all exploited and given supernormal signs in cultural activity. A particular Dogon mask evokes a curious reaction of fear, loathing, and desire: why? Because it activates several sign systems at once with a supernormal intensity. Arguably, much of the human art of psychologizing, is the supernormal stimulus of social signaling for the purpose of acquiring power over the target: psychology is a weapon, and the urform of the "syndrome" is *the mask*. A mask calls a spirit: and a spirit is, from this perspective, nothing but the coagulation of excess social response into a *relatable* form. Tribal apes deal with "persons", therefore anything which must be dealt with, must first *become a persona*: thus each spirit has its mask, and each mask has its power of supernormal stimulus. We are, as the most teachable techno-ape, not only "freed" from instinct, and thus prone to instinctual incoherence and thus anxiety and thus neurosis, but gifted and cursed therefore with an *excess of instinctual energy without unambiguous aim*: we become both *more excitable* via supernormal stimulus because our instincts are weakened in aim, and *numb* to stimulation because we do it to each other almost incessantly. In postagricultural civilization the acceleration becomes unsustainable. Every postagricultural culture is an orgy of mounting overstimulation and an arms race between the subtle arts invoking a set of advantageous responses, and the exhaustion of those responses: the pyramids grew in size until they no longer inspired the Egyptians to be Egyptian. Greek austerity eventually was not a reprieve from Greek sensuality, but merely boring, because they were no longer vivacious enough to crave the contrast.

- 89 -

Analogy for engineers, concerning the value of meditative practice designed to optimize low-level function: it doesn't matter how good you think your software is, if it's running on a faulty network. In order to *act* smarter, which is to *become* smarter, you must improve the performance of the entire biological computational network: that means the peripheral nervous system, the somatosensory, the sympathetic and parasympathetic – everything which most of us take for granted. Most of what holds back my typical reader is not a dearth of intelligence – generally a surplus of cleverness reigns – but a lack of systemic integration. High degrees of inefficiency and competition between overlapping systems is the norm: most engineers and scientists I've known utilize only a small fraction of their potential intelligence, but with a kind of hypertrophic lopsided metabolism. They walk around with their craniums precariously suspended, like a zeppelin trailing so many fraying ropes. But these are never the most intelligent, nor creative, nor motivated agents of our world. They bury their heads like ticks into a specialization and hope thereby to have hidden the rest of their frailty. My path points the other way: towards a bottomup optimization, in which outcomes are stochastic but determinative, gradual but inevitable, unannounced but dramatic, unconscious but cybernetic, diffuse but immersive, like the imperative unity of a choir compared to the droning of a lecturer, like the ominous receding of an ocean wave, it may seem like less before it overwhelms with more.

- 90 -

Why is everyone so solely focused on the brain? Because its performance seems to be correlated with *social status*: the peripheral and somatosensory systems I'm speaking of are generally neglected in neuroscience because they seem to be of low social value, and thus the modern human creature assumes they must be of low *functional value* in the development of intelligence, and even of low *ontological value* for the the question of what it means to be alive. The commonality of the wish to be a mere brain-in-a-jar, is only growing in the 21st century – outside of the rare professional athletic career, a finely tuned proprioception seems to grant very little social status: no one cares in a typical job interview how mean your pingpong is.

But the reality of high performers does not reflect this assumption: the topshelf corporate lawyer who gets up at 5am for her daily intensive pilates workout, doesn't do it so much for the muscle tone but because it calibrates her peripheral nervous system such that she remains sharp-as-a-tack and hypervigilant concerning her best interests. It's not just her "brain" that makes her formidable, but the asynchronous orthogonal operation of multiple systems near criticality which gives her that uncanny memory and attention to detail. In thermodynamic terms, a body like this is capable of sustained performance at phase thresholds requiring high dissipative throughput. This is possible because she stresses all her systems regularly, such that when realtime performance require their function, they already reside in a raised basin of attraction – a learned minimum which is near the global maximum: a high performer like this is generally both relaxed and fully poised, not tense yet never slack, unpredictable in the details yet fully reliable. But it is the synergetic effects of multiple overlapping systems which make for brilliance: this is possible only when each relevant system is "happy", which means not generating noisy errors. Her emotions don't get in they way of her intellect, which doesn't overload the capacities of her body. Obviously I have different means and ends than this manifestation of human potential, but it's illustrative to see the principles hold nonetheless. "Spiritual" development is everywhere when you learn to see it.

- 91 -

They've been going about it backwards – these artisans of artificial intelligence. Symbolic reasoning is not the *prerequisite* to intelligent behavior, it is the *consequence* of coherent sensorimotor preprocessing.

Every object choice is a *symbolic object choice*, only made possible by rigorously refined and *amplified* data propagation. A "symbol" is what remains of a signal after iterative filtering. Therefore stochastic resonance is likely at play, everywhere perception is at work: an optimum noise complement makes relevance discoverable, because otherwise a sensory field could not anticipate vital signals fast enough. Thus it is that sensory deprivation will yield up the baseline hallucination which forms the bedrock of all perception: this is how dreaming was born, and why it became yet another opportunity to refine the same processes of sensory-symbolic generation... Freud was absolutely correct that the dreamstate is characterized by a tightly packed vector field of symbolic valence, but wrong that its traversal is motivated by mere "wish": the organism wastes no opportunity to rehearse its primary challenge – which is to navigate its own world of signs, to swim confidently through its niche for a lifetime, exploring every cranny of its ancestral mind.

Likewise, we only know what anything "means" because of the almost incorruptible coherence of the lower levels of sensory processing: we are *presented* with meaning, and then ascribe to the process a mysterious agency like "intention". But almost all conscious reasoning is *ex post facto* reasoning: we fabricate memorable tales to tell each other, and only as an afterthought do we spin the yarn to *ourselves*. The "I" is a consequence of sociality and was invented long after the You and the We. The individual appears only after the tribe is millennia old.

THE MORAVEC PARADOX IS NOT PARADOXICAL

LIMITED INTELLIGENCE OR INTELLIGENT LIMITATION

- 92 -

An IQ score measures the ability to succeed in the currently dominant human societies – no more, no less. If you define intelligence as successful behavior in a given environment and test for those traits by simulating the determinative moments of that environment, you can predict socioeconomic status and claim therefore to have discovered "intelligence itself".

The sedentary, tame, anxious-yet-ambitious modern human being asks itself, "What's intelligent?" It answers: "To acquire status". It then asks, "What do you need to acquire status?" And answers, in this order: "Verbal dexterity in a standardized language, fluency in dominant social expectation, excellent short-term memory and concentration, basic arithmetic, unhesitating response, and finally a little 2-dimensional visual reasoning sufficient to fill out taxforms, navigate a streetgrid, and find the bathroom." Thus is the IQ test born.

What's left out?

¬ Longterm planning, determination, willpower.

¬ Imagination, creativity, humor.

¬ Holistic physical coordination. Proprioceptive accuracy in 3 dimensions.

- Rhythm: fine muscular control, breath entrainment, mood modulation. As important to the athleticism of Paleolithic hunting as to music.

- Empathy: the ability to anticipate and detect uncommunicated affect.

- Emotional equilibrium: the ability not to react, as a precursor to gathering yet more subtle data.

- Courage: or the ability to minimize risk by assuming risk.

- Honesty and loyalty: or the ability to create longterm social value through renunciation of trivial gains.

- Sagacity: or the ability to synthesize experience into dense, memorable, and transmissible knowledge.

Seeing this list, actually the number of core human virtues *not measured* by our "intelligence quotient" is rather startling, isn't it? Again, the IQ test measures *the likelihood of success in currently dominant social structures* – and admittedly, it does this quite well.

Moreover, what everyone forgets is being tested, is this: the *willingness to undergo an intelligence test*. In other words, the degree to which the human creature is already acclimated to the expectations inherent in the test itself. Or, the degree to which any given dope is already suited to an office job.

- 93 -

But the most measurably successful types are not those who score the highest IQ: CEOs, financiers, lawyers, politicians and other exquisite prostitutes are typically slightly above average but no higher – otherwise known as that sweet spot where a very small advantage is immensely reinforced because the Dunning-Kruger effect still holds. Just smart enough to feel smart, but not nearly smart enough to feel dumb. My guess is that the most successful in our world rarely exceed the first standard deviation at 115. But what *do* they possess? The ability *to tell lies under pressure*, to sustain cognitive dissonance, to exude charm and the aura of success: if you designed a test for *these* qualities, I'd wager you'd discover both those who succeed wildly and those who fail spectacularly.

If you look carefully at the highest IQ scorers, aren't they boring unimaginative people? Perhaps very good employees? Agreeable chemists? Reliable mortgagees? What does the IQ test measure but one's likelihood to make a *diligent clerk*? Isn't that rather unsurprising, given *who exactly* designs such tests and therefore would like to imagine themselves as the highest type and destiny?

\- 94 -

Once again I find myself in an odd margin of the intelligentsia: I despise egalitarian nonsense, but I also despise the smug agencies of civilizational conformity which would elevate current humanity and only current humanity above all other forms of life. Pride concerning who one is relative to the rest of the slobs is valid. But humility concerning one's place in nature and therefore one's identity with all of humanity is equally necessary. We are excellent, but we are also merely another animal among many: we are not the most beautiful, nor the noblest, nor the only one that reasons, nor the only one that communicates, nor the only one that makes war and peace, nor the only one with cities and agriculture and slavery, nor the only one with cultural knowledge, nor the only one that feels love and hate, nor hardly any of the criteria of exclusion which we have employed in the past when we wanted to say, "We are not animal". What distinguishes us therefore, if not "intelligence"? We are the *most wicked ape*, and perhaps the most *problematic* earthling.

- 95 -

In other words, there are many aspects of general intelligence which must be almost impossible to quantify. Intelligence in all its manifold operations is difficult to measure because it is so contextually sensitive. Stupidity, however, is easily detected: thus I insist that the IQ score is statistically reliable not because it measures "general intelligence", but because it demonstrates *distance from easily detectable stupidity*. The most intelligent creatures are *diverse* and *eccentric*: it's very difficult to predict how and when very bright children will demonstrate their gifts. Stupid behavior, however, is all the same and thus makes for an excellent statistical bullseye: slow, numb, loud, redundant, lazy, unmotivated, unwilling.

Notice again, that in any behavioral test, much of what's being tested is the *willingness to undergo testing*: the subject must be convinced there's something in it for him – he must already be invested in the ritualized power relation and seek therefore to perpetuate it with his own body. There are a thousand and one idiotic experiments concerning the intelligence of animals, in which a petty clerk assumes his experiment is so self-evidently important that any animal which shows no interest in cooperating must therefore be stupid: when perhaps it is we human creatures who have lost our way, forgotten what matters in this earthly plane, and often enough behave much stupider, coarser, and meaner than our core instincts would require... Half an education and a magic robe of science is a very dangerous thing to gift to an ape.

- 96 -

If you designed a test for success in a Paleolithic hunting tribe, many of the first shall be last and the last shall be first. Most of us would just fall off the back end of the scale like a drunk on a hayride.

- 97 -

What I find particularly pernicious about the current controversy surrounding IQ distribution in various populations, is that it establishes a misleading dichotomy between the histrionics of egalitarian social constructivism and the obscenities of crusty white-supremacists. As though there were no alternative: either be a moral actor and deny biological determinism, or imply that European political ascendancy is the will of God. The first has no intellectual conscience and seeks merely to escape moral censure, the other has no historical perspective and seeks to escape white guilt by confirming its necessity.

- 98 -

Concerning the "white-black gap", what's rarely mentioned is that subsaharan African genetics are in fact *the most diverse* on the planet: just because everyone's roughly a shade of brown does not constitute a "race". The fact is that the kind of *ritual* the IQ test represents is not something a subsaharan individual is culturally prepared for.

Here's a nice thought experiment: it's known that northeast Asians score the highest of anyone. It's also known that these same people and Native Americans are genetically quite similar – the crossing of the Bering Sea was not very long ago: yet Inuit and other tribes score on average much lower than Koreans, for example. Again, I contend that the difference is not wealth and privilege, but *cultural entrainment*: high-scoring North Koreans represent some of the most culturally disciplined people on the planet, fully prepared to undergo a meaningless bureaucratic test under extreme psychological pressure – the Confucian state exam will probably remain the longest-running clerical test in history.

Studying behavioral characteristics as though they could be separated from cultural entrainment remains extremely dubious.

Human intelligence varies widely *between individuals*, not between groups: as a whole humanity is remarkably uniform given our worldwide distribution, but this is because our common ancestor was not sufficiently distant to make much difference...

- 99 -

"Biological determinism" is nonetheless the correct point of departure. There is a definite possibility of a world in which meaningful human races *could* exist – from the perspective of avoiding moral complications, we're merely fortunate that race is largely moot. But there is no reason to assume that castes are impossible. Indeed the rivalries and occasional interbreeding between the various hominid species of the past was a world in which race mattered a great deal: our ancestors struggled with competing races for hundreds of millennia while stealing a few genes here and there. There are fragments of Neanderthal, Denisovan, and Homo erectus among us even now.

It's been my intuition for years now, that *Homo sapiens* was the *not the most sapient*, nor the bravest, nor the noblest: he was the *most treacherous*. The best liars, the best craftsmen of evil snares, the best cheating merchants, the sneakiest thieves of technology: our kin. There is good evidence that it was Neanderthal who invented and perfected the hunting techniques we depended upon for hundreds of thousands of years, that the Neanderthal used herbal medicine, and that they had larger brains – it's a scandalous secret that human brain size has been *decreasing* for the last 20,000 years.

- 100 -

The fact that the psychometric of general intelligence, "g", seems to correspond best to *verbal comprehension*, only proves that such tests are centered around the question of verbal intelligence. It would be just as easy to to design a test which preferred 3-dimensional reasoning as the pinnacle of general intelligence: one would only have to be implicitly testing that reasoning in every other category. Moreover if *g* corresponds with socioeconomic status, then again, all that is proved is that these tests successfully predict likelihood of socioeconomic success for the average dope.

Human language is the most recent, most visible, and therefore *least profound* manifestation of "biosemiotics". It is only the *intelligence embodied* in everything else which surrounds this incessantly talking head, which provides the illusion that the talking does the steering and effectuation. Precisely the same prejudice is at work in the vogue fascination with "the brain": the human creature wants to be judged for the last 0.001% difference between himself and his neighbor, rather than the 99% of what he shares with the chimpanzee – how much less the majority of what he shares with the shrew?

- 101 -

Limited intelligence or intelligent limitation: one of the worst pieces of anthropocentrism, or what I call "arrogant apethink", is the assumption that any creature displaying a less complex feature of some trait we value as part of our intelligence must to that precise degree be less intelligent than us – as though there were only one kind of intelligence, and as though we embodied all possible features of intelligence. For example, birds sing – but they do not speak. Some of them sing complex songs, but because there is no discernible syntax and no obvious displacement of information, we shrug and say it is only a degenerate form of communication, ignoring the fact that it is precisely song that underlies oral culture and thus human culture as it has been for most of our history. The truth is that a successful living creature maximizes its abilities *within its particular adaptive zone*: limitation can be as adaptive as ability. The most successful animals are masters of their art – and in any given art the limitations are the stage upon which the mastery plays. Would it be "better" for birds if they talked as we do? Would they thereby grow more intelligent, or perhaps less intelligent *as birds*? Is it an indissoluble element of being a bird to communicate precisely the way they do? Could it be the *totality* of each phyla which lends it staying power and adaptive value? Isn't that a kind of "intelligent limitation"? Every ability is also a liability. A bird that could fluently speak a human language would most likely not be viable – it would have strayed from the birdway, and become only a monstrosity and a helpless pet.

THE BIRTH OF PURPOSE

- 102 -

What is biological intelligence? The question should be pursued not only in terms of competence, but creativity. Intelligence does not merely outperform other alternatives, it *generates solutions* in a richly textured unpredictable environment. The ability to generate *contextually dependent solutions* through the interaction of a small set of modular behaviors defined by metastabilities in statespace: ethology used to call this "instinct", and this is where I believe we should begin.

- 103 -

First, a little historical overview from the perspective of a psychoanalyst:

The Freudian, Lacanian, and poststructuralist theories of unconscious process are much more compatible with mathematical modeling than generally assumed. It was Freud and Levi-Strauss who first designed a theory of symbolic motivation compatible with the *vector field*: existing neural network models and NLP techniques actually owe a great deal to early 20th century psycholinguistics. The inventor of the perceptron and thus the forerunner of all connectionist models, Frank Rosenblatt, was himself a psychologist. W. Ross Ashby, the first to write systematically about cybernetics in the 1950s, was also. The still fundamental "McCulloch–Pitts" neuronal model was built from the collaboration of a logician and a psychologist.

But what's still missing is an integration between neural network principles and the rigorous formulation of *unconscious mechanics*: these mechanics themselves have been known to psychoanalysis for more than a century, proven in a thousand contexts, and yet remain largely unexploited in any computational context. Partly this is due to the exotic flavor continental philosophy still has for the Anglo-American sphere in which nearly all AI and neuropsychology has taken place, and partly this is due to the inevitable resistance which any psychology of the unconscious encounters: part of how unconscious process secures efficiency, is denial of its existence.

- 104 -

One of the insights of the mathematician Norbert Wiener, is to have conceived of a signal as a special kind of noise. The psychological analogue can help explain what I mean by "unconscious mechanics": namely that conscious process is largely constituted by *redundant marking*. A conscious topology is coherent but informatively sparse. Since the Romantics of the 1800s, the realm of unconscious process as been popularly characterized as lawless, arbitrary, and irrational. Nearly the reverse is the case: what is unconscious can afford to remain so, because it's metabolically optimized. What we're conscious of, we do awkwardly, slowly, expensively. It was one of the important results of chaos theory in the 1990s to emphasize that chaos should not be properly considered "random", but a state of excitation produced by the competition between many potential kinds of order, such that the total statespace of a given system lacks easily detected structure. Similarly, biological noise only appears random because the cumulative amplitude reflects too many trajectories simultaneously to be formulated as a linear function. The great biologist E.O. Wilson had a similar realization while observing his ant colonies. At low states of excitation, the collective behavior of Hymenoptera can appear futile and aimless: a single ant may pick up a pebble in the vicinity of the entrance, move it a few inches, drop it, walk in a circle, pick it up again only to move it back to its original position. But this is the stochastic foundation from which purposive behavior is built: it requires many nonlinear interactions between many relatively free agents acting within the strict restraints of a finite system, to produce holistic "agential" behavior. It was the principal insight of cybernetics that in a system's search for homeostasis, purposive behavior is born. Biological systems are inherently dissipative, and therefore designed to withstand enormous energetic throughput such that they are capable of *exploiting the fallback* in generating organized purposive behavior. This kind of self-organizing novelty cannot be obtained through linear guidance, but rather must be obtained through the system's own exploration of its critical moments.

- 105 -

An unconscious topology such as we experience it in psychoanalysis, is above all richly connected and *overdetermined*: any one effect has many causes. A hypothesis reliant on a *bijective* relation between intelligent result and uniquely determined mechanic will always fail. The systems we're discussing are fully deterministic in shortterm pathfinding yet stochastic in longterm evolution – their statespace trails have rational causes but are extremely difficult to predict: my argument is that the same characterization applies to unconscious process, because although the semantic valences with which it deals are highly ordered, their extreme connectedness makes surprises the rule rather than the exception. In the classic example of displacement and the element of the uncanny in dreams, the object choice which serves to obscure the repressed content often has the appearance of the comically arbitrary, but upon closer examination was always assigned its role due to a maximal connectivity to the original repressed ideation: Freud's phrase "Wortbrücke" was apt and prophetic, considering the current direction of natural language processing. Neurosis forms as a locus of high connective throughput, which encapsulates a region of the repressed like scar tissue. Arguably, even ancestral instinctual centers in an unconscious topology produce similar distortion effects. This is what one of the last great naturalists, the ethologist Nils Tinbergen, also incidentally termed "displacement": when the primary instinctual path is blocked, such as the consummatory act of a mating ritual, other peripheral behaviors become activated through displacement. A male goose frustrated in his courtship will instead peck at the ground furiously. In this one easily pictured example, we see both the origin for the capacity for repression – namely the halting of instinctual discharge in a social context – and the evidence of the protolinguistic relationship at work in any given displacement. Namely that what the male goose wants to do in the fulfillment of his duties is symbolically enacted in the way he pecks at the ground – a substitute behavior which is nonetheless "understood" by all.

\- 106 -

Everything I've said could at least be employed in making natural language processing more interesting than it is now. In my opinion, its output is largely boring, only vaguely plausible, but not valid. It's merely the statistical average of a great many texts organized into a vast lookup table: there is no meaning because there is no driving force, there is no driving force because there is no theory of hermeneutics behind the modeling, nor any model of a constrained system undergoing energetic stress. The wonders of GPT are obtained through bruteforce combinatorial computation: any given word's likely occurrence in the vicinity of any given word. It's my contention that this has nothing to do with biological intelligence nor what's valuable about human communication: which is why its output reads like a mediocre 10th grader trying to sound smarter than they are via plagiarism and a thesaurus. We still have an opportunity of doing something more significant than improving on the latest chatbot: as a scholar and psychologist, this interests me much more, despite the fact that the big cash money wad probably lies in the direction of producing vapid results for the sake of credulous crowds. The 21st century bigtop and sideline freakshow takes place digitally, but the crushing stupidity of the majority remains the same. Yet the activity of the field tells me that I'm not the only one thinking this way, and that someone will implement these ideas in the near future...

PART III:
APPERCEPTION

INTELLIGENCE SLEEPS IN YOUR FINGERTIPS

- 107 -

The benzene ring was discovered in hypnagogia, as a talented chemist stared into a fire: he saw an ouroboros as the solution. Hypnagogic thinking tends to be characterized above all by recursivity: self-referent metaphors, cycles spawned by cycles, a compulsive thought which cannot unravel because its traversal engenders the compulsion. But this is no mere neurological errata, no mere footnote in the history of affliction: my contention is that all deeper reasoning is recursive by nature, because a whole cannot actually be conceived, but a *scale-invariant repetition* can be inferred. What is a whole? Not merely a bounded surface, not merely an "object" and thus an erasure of depth: that is the influence of the symbolic reasoning of narrative consciousness, and not the full inheritance of our unconscious intelligence. When we intuit, we conceive of wholes. A whole implies stability, which implies circular causality: even a dissipative system as simple as a dustdevil testifies to a circularity of cause and effect. An imbalance of force constrained to a boundary such that order emerges: tension, compression, and critical resistance are involved when we see a "whole". Both topdown and bottomup: the ancient habit of seeing "spirits" in everything, was not as neurotic as we like to imagine. There is a deep rationality to animistic thinking: what is a *spirit* but a generative function permeating a field? A scale invariance producing the illusion of personage, and thus *requiring a mask* both as honor and limit? The nomadic attitude toward spirits was generally very wary: they are infectious by nature, because in our language a spirit is *a recursion unconsciously perceived* – whether our ancestors spent too many millennia hallucinating wildly under the influence of this newly liberated oversized cranium of ours, prematurely peeking into spatiotemporal secrets which we have only begun to soberly examine, is a problem I to leave to the reader. Our immediate question is: *can we responsibly hallucinate?*

- 108 -

But what is the productive imagination, if not *responsible hallucination*? To maintain an intellectual conscience, which means to "show one's work", to be able to converse, to engage wholeheartedly in the ongoing dialogue, to be able to state reasons and elaborate and build something viable over time: just that much *requires* already a productive imagination... Therefore we're led to the conclusion that an intellectual conscience cannot be nurtured, exhibited, or even *exist* without its counterpart: a wild enthralling hallucinatory power. Everything else is just monotony, mendacious conformity, and cowering complicitude which wants to pass off its *inability to create* as virtue: my point is that good science happens in the tension between the excursions of hallucinatory elaboration and the reductions of articulate dialogue. Therefore we, who are interested in developing our powers of knowledge and foreknowledge, who cannot resist the temptations of the bleeding edge, should develop not only a reductionistic habit of expression but a deepening ability of controlled, *willed hallucination*: for there is no other way to perceive.

- 109 -

There is an intimate relation between topology and multivariate causal reasoning: this leads here, that leads there. It's been my impression that the mathematical power of "statespace" is nothing other than the cerebellar geometric imagination applied to the time domain, and thus the hominid talent for *toolcraft* is invoked. Consider what it requires to make a fine handaxe from chalcedony: one must constantly conceive of the motion of the tool in time, along with the transmission of lines of force along the semicrystalline structure one has in one's hands at that moment, examining every ridge and valley for weakness and cohesion, channeling force from the leading edge to the grip without yielding to any local congestion, such that the relative brittleness of the material is maximally attenuated while its hardness and inelasticity is maximally exploited. Take the time to examine a genuine Clovis point, you will find in the most masterful pieces nothing less than this topological sophistication and indeed more – for there are many other considerations than I've mentioned: craft is not something that yields up its secrets to language so easily.

I have the privilege to live in an area of high desert where obsidian arrowheads litter the ground, and one cannot know if any one piece is 200 or 14,000 years old. What we're discussing is not a trivial task: try a little flint-knapping if you believe the stoneage to be "primitive", and that word to signify "easy" or "stupid".

- 110 -

What therefore is the importance of *Maxwell's surface*? Why has topology become so important in modern science?

Because we are finally undoing some of the willful stupidity of postaxial thinking: namely the famous obtusery surrounding *linear causal reasoning*. We are accomplishing this by a graduated return to the nonlinear, multivariate, and overdetermined causal reasoning inherited from our hindbrain.

In this context, "linear" represents a cluster of philosophical prejudices surrounding causality:

1. Unidirectional
2. Inharmonic
3. Uniquely determined

We will have to examine each of these in turn. But it is enough to say that *social posturing* is the root of the worst causal thinking with which our sciences are still rife: the ape only *pretends* to be so dumb as to believe in a singular cause for any event, and only when he finds it socially advantageous – the problem is that it's so useful as to be practically indispensable. Thus is blame mobilized, thus is maintenance and adjustment of social hierarchy made fluid, thus is *ritual* made possible which achieves very real consequences, and thus is *histrionic certainty* generated which is so necessary to get a loosely allied troop of primates to do anything at all. *Our sciences are still ritual*: look behind the veil of the more ambitious journal articles, and you will discover many reasons to subtly fabricate, to gingerly omit, to cloyingly exaggerate, to *pretend to be stupider than one is* – in the hope that your audience has reason enough to join you.

Linear thinking is social and histrionic in origin: it underlies all hysteria, all as-if, all posturing for clandestine benefit, all disguised social maneuvering. The more panicked and lonely the ape becomes, the more it resorts to a pretense of obtuse linear reasoning – even while it practices an even more opportunistic and multivariate unconscious calculation. There is an intimate relationship between the myth of agency, the superstition of free will, the soul, and the assignment of blame on one hand, and the witch-hunt, ritual exclusion, postaxial religion, and the moral disease on the other... But we've been here before, and this time the path will only strafe these lessons.

- III -

Tactile, visual, and auditory processing is the root of sophisticated causal reasoning: the reification of shape, the assignment of locale, the inspection of invariance under transformation, the ape specialty in *the intelligence of the hands*.

- 112 -

Wenn wir einen Gegenstand abtasten, erteilen wir seiner Oberfläche mit Hilfe unserer tastenden Finger ein feines Ortemosaik.

As we feel out an object, we deliver a fine mosaic of place upon its surface with the touch of our finger.

Uexküll, *Streifzüge durch die Umwelten von Tieren und Menschen*, §1.A

I have examined enough arrowheads to discern the degree of craftsmanship involved in any one: some of them are masterworks, and a signature of the quality of mind and body which created it. Some of them are witness to immense patience, care, and observation. The balance between improvisation and exactitude, the even spacing between strikes, the gentle rotation of angle of incidence, and the final sign of mastery: the logarithmic slope from center of mass to edge. But the best way to see this is not with the eyes: you must put away your eyes and use your fingertips. The surface should feel like rippling water, like the sound of rustling silk, like the curvature of a blade of grass: don't doubt that the knowledge of these formulae resides in the human fingers – don't doubt that the core of your intelligence dwells right there, caught in the whorl of the human fingerprint.

THE UNDULATING SURFACE OF PERCEPTION

\- 113 -

One of the most important concepts from set theory, which informs my approach to meditation and consciousness, is *surjection*. Surjection describes a transformation between two sets, in which many points map to single points: a many-to-one relation, a reduction of complexity. In both psychoanalysis and the fundamental causation theories of physics, this is known as "overdetermination". It also happens to be the only sane model for the relation between unconscious and conscious process: a conscious topology is densely packed, multivalent, uncanny. People always mean more than they explicitly say: just as any given apperception is a radical simplification. When we meditate, we are narrowing our focus and amplifying the recursive function of apperceptive consciousness, so that we get a mysteriously undulating surface which always implies more than it specifies: much of the art of insight is to pay more attention to the undulation than the surface – therein are the latent variables, and thus the interesting data. Just so, in narrative consciousness it is often the sequence of words, the omissions, the phrasing, the timing, the context which reveals much stronger valences of intention than any explicit semantic digestion – the much lauded "natural language processing" of machine learning, for example, consistently fails to map sentiment on any dimension greater than one: as though emotional valence could ever conceivably be so shallow as to fit through a "positivity filter"...

When we meditate, we watch this undulating surface of perceptions, fragmentary thoughts, urges, itches, memories, unfinished arguments: it's like sitting on a street corner of Manhattan observing pedestrian traffic. There's two sentences of one side of a tense phone conversation, there's the smirking suit thinking of his bank account, there's the 20-something hottie enjoying the bubble of distraction she creates, there's the dog walker looking bored, there's the tourist looking stupid... Each case represents the *media res* of a long story, a mere fragment which is legible only to the degree that one's imagination and experience transforms it into a probable history: just so, every distracting thought in meditative practice is a rambling story, which has its own valid reasons for being. Proper technique neither indulges the talkative mind, nor suppresses it: we have to learn to grow up and stop indulging every urge, yet we also have to respect ourselves enough to forgive ourselves for being so compulsive, frayed, pathetic. Meditation is the great confrontation with what we really are: an anxious wreck, an obsessive little shit, a resentful arse, a rebel without a cause, an arrogant bore, a tired drone, a desperate loner – we have all been these things and more. None of us is alone in the *cause célèbre* of modern wretchedness: the proliferation of consolations for chronic anxiety should teach us that much. But rather than make it worse through *endless consolation* – indulgent psychotherapy, spa treatments for the soul, pharmacological narcosis, guilt-laden consumerism smeared with political pretension, the *cargo cult narcissism* of social media – we recommend rather the old-fashioned approach: *poised discipline.* A basin of attraction in critical emergent systems has the odd ability to reverse the attraction-repulsion factor as distance varies: a strange attractor should be properly named an "attractor-repulsor". At close range, a critical moment behaves like a steep ridge, discarding approaching trajectories. At wide range, the same moment draws in distant paths: this has been my experience with disciplines and excellence in all fields. The closer you come to mastery, the further it recedes and the steeper the path. Meditative discipline as I conceive of it, is the practice of excellence distilled to its core: we just sit there, breathing, trying to not be such an impulsive coward. The rewards of the practice therefore, apply in every scenario: everything the body does can be construed as a function of emergence through constraint, metabolic efficiency, dissipative flow. The more advanced students of philosophy will probably recognize in my writing method, more than a hint of the Nietzschean pedagogy: the trick is to allow playfulness even while shepherding the flock toward some distant goal. Nothing else yields the feeling of a genuine conversation with a genuine soul: old Fritz composed most of his best material while walking, and it shows. Beware all thought born of "Sitzfleisch", he says: odd then, that I'm comparing *sitting meditation* to Nietzschean excursions? But this is my point: proper meditation is a virtual journey, because we cannot stop the spirit from wanting to wander, we can only learn to compose it, to gently contain it, to amplify excursion into recursion, to wait patiently for the return, to watch ridge become valley, to trust the undulations of a neurochemical bath and learn precision and poise in place of stasis and fear. The surface of consciousness represents *potential*, not result.

THIS BRILLIANT QUAKING BODY

- 114 -

Why topology? Because it aids in the visualization of *invariance under transformation*: much of what's interesting in topological reasoning, are the surprising ways that the properties of Euclidean space can be subtracted, without losing feasibility nor mathematical rigor. In other words, topology, graph theory, and statespace reasoning aid in the fruitful visualization of neuromorphic experience: what it feels like to be scintillatingly alive, is something like a flashing tapestry folding and ribboning, taking shape and losing it again, always hinting at quasicrystalline regularities but dissolving when the perspective shifts and reveals variance. What's truly invariant in meditative experience is the structure of the neural body itself: not "reality itself", nor "the noumenal", nor any other way of saying "the mind of God". Moreover it's not "mind" which is revealed, but the emergent attractors of a structured neurochemical bath in this brilliant quaking body. What's the difference? "Mind itself" does not exist: the body is *defined* by existence. It is the unfathomability of the existence of this intelligent unconscious body, which the postaxial philosophers fled: they could not handle the psychotropic side-effects of the abyss of genuine mouthless rationality, genuine biological computation surging through us and over our heads like an impossible tidal wave, the convergence of unthinkable indifference and the most personal forces, we ourselves. This is what deep meditation will teach you to withstand: the abyss of light at the center of your being.

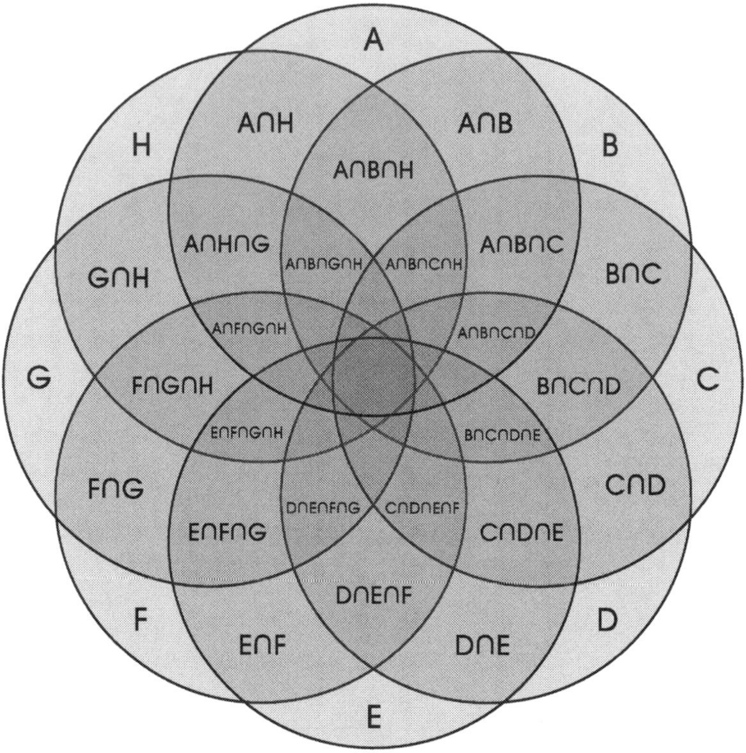

Intersection of sets as an abstraction of location

- 115 -

> The specific attraction and in a large part the significance of topology lies in the fact that its most important questions and theorems have an immediate intuitive content and thus teach us in a direct way about space, which appears as the place in which continuous processes occur.
>
> P. Alexandroff, *Elementary Concepts of Topology*, §1.1

What is topology? It is *an abstraction of space*. It is such a radical abstraction, that it reveals the intertwining of the concepts *shape* and *space*: what is space but a continuity of potential points? What is shape but a surface, and what is surface but a bounded set of points?

General topology achieves this abstraction by discarding the notion of *distance*, and replacing it with the concept of *continuity*. A little familiarity with set theory is required, but like most powerful notions in mathematics, it's not the complexity which is difficult but the austere *simplicity*.

- 116 -

The open sets of topology have no "location" in the usual sense. They are solely defined by their membership roster of points, the constituency of which reflects location but no longer contains a localizing function except through their interaction. It is the *interaction* of these subsets which "builds back" the original shape: a topology is therefore a kind of abstraction of location which can pass through a tight wormhole of minimal representation without losing the essential datum: an approximation of *where*, given any point which is a member of some collection of these subsets.

> Der elementare Baustein des Tastraumes ist keine Bewegungsgröße wie der Richtungsschritt, sondern

> eine feststehende, nämlich der Ort. Auch der Ort verdankt sein Dasein einem Merkzeichen des Subjektes und ist kein an den Stoff der Umgebung gebundenes Gebilde.
>
> The fundamental building block of tactile space is not momentum as in the directional step, but a fixed one, namely place. Place also owes its existence to a perception-sign belonging to the subject and is not a structure dependent on the matter of its surroundings.
>
> Uexküll, *Streifzüge durch die Umwelten von Tieren und Menschen*, §1.B

- 117 -

Think of the familiar Cartesian coordinate plane. The important thing to understand about every metric space like this, is that the location of points is defined as units of displacement from a *privileged zero point*: in topology and affine space, there is no such thing. We're left instead with *relative location*, defined in topology as subset membership and in affine space as relative displacement vectors.

- 118 -

The reductionistic power of *affine structure*, is to have replaced the concept of measurable distance with the concept of a *ratio of displacement*.

Likewise, the powerful abstraction at work in topology is to have replaced the notion of an axis of orthogonal coordinates which define any given point, with the notion of *continuity between sets of points*. To know the location of a given point, one finds the "neighborhood" sets including that point. This is somewhat like triangulation, except this is a collection of strictly ranged binary detectors, arranged around the point of interest like a lotus...

Could it possibly be that the phyllotactic spirals so common in plants of all kinds, has more than a little to do with the concentric circles and spirals of visual processing artifacts, as a result of *optimal packing* both in the retinal-cortical mapping and the buds of a growing stem? Heinrich Klüver, stuffing himself with mescaline in the 1920s, thought as much: moreover could such "form constants" be the result of needing to *readily hallucinate* that which nature produces? Nature must be legible for the nomad. In other words, the longtime fascination with the lotus is not merely about the "transcendence" of mud: self-organizing dynamics in neural experience is one of the many portals *back to earth*. Such a portal is visible in something as humble as the sunflower: we want *geometria* to be the science of "untamed earth", just as Mandelbrot dreamt...

- 119 -

What I've just said is confirmed by that special night I spent in Huichol territory among the cacti of the high desert, when I kept seeing the face of a Peyote button written in concentric circles of the stars... What I'm getting at, is that we must steal back *sacred geometry* from the weak-minded hippies, wash it clean of self-indulgence with mathematical rigor, blow off the stink of moral posturing with our dragon's breath of unremitting critique, and make it warrior's food again: tantalizing rational correspondences between the branches of axis mundi are supposed to embolden us to act as though our ancestors were watching our every move, as though the very rocks were "grandfather" – and since we share so much, since the frontier between living and dead is not so distinct in the light of nonlinear self-organizing dynamics, who's to say whether we have a right to claim as ancestor everything that seems to speak the hidden language of our most repressed and liminal experience – and what's math but another hidden language?

- 120 -

In other words, the *worst of ape-nature* stands between us and a more genuine spiritual life. It's probably absurd that we find such intellectual gymnastics necessary just to feel a little more at home on planet earth. It's probably funny that so many skillful rhetorical turns seem necessary just to avoid the suspicion that we're full of shit: but we find the road littered with wrecks and landmines. Everything which prefers shallow social strategy, shortlived moral posturing, frail wishful conclusions planted in the ground like monuments: these are our obstacles, and a broad historical education at first seems to *increase* the danger of spiritual paralysis, rather than familiarize us with follies. We alienated, overeducated superfluities of the 21st century *want to go home*: but if we're merely mythmaking, with our crisp psychology and antimetaphysical mathematical trance, isn't mythcraft also within the *core human adaptive package*? To make excursions to the limits of knowledge, looking for medicine power: are we doing anything different? Homo sapiens is cursed with freedom from instinct, and in place of the unquestionable certainty of instinctual imperative, he must contort his half-insane dreaming mind into something edifying... An inspired madness that compels rational behavior: is there a better definition of the divinatory art?

- 121 -

What we lose in precision, we gain in abstraction. And oddly enough, this is closer to the reality of biological computation than a Euclidean distance function ever was: most processes of mapping and path traversal yield to topological analysis much more naturally.

- 122 -

The expressive power of set theory seems to come from the opportunity to perform rigorous *binary operations* over arbitrarily large datasets: an element is either in the set, or not. Thus an array of subsets over some set of points, provides a neat abstraction of the concept of metric – namely that a region or "open ball" is replaced with the notion of subset, and you're left with the ability to perform binary operations like intersection, which replaces the more familiar arithmetic calculations of coordinate values. And it is the union of subsets which creates an interwoven matrix constituting a topology: which is only rendered back into something like metric space, when the collection of subsets is indexed and iterated, and the overlapping membership and exclusion of each point pushes these subsets into their proper place – they mutually align and form a fabric. *Location* of a point is not contained in any given datum, which allows for extreme abstraction and economy of notation, but the *membership distribution* itself contains the shape of the space thus analyzed.

- 123 -

I had a hypnagogic vision related to this subject. Mathematicians don't like to give away the simplicity of their ideas willingly, so I've found it often requires a vision before I feel I have a right to them: the mutually reinforcing coexistence of rigorous definition and poetic elaboration is not common knowledge... After having spent a few hours hitting my head against an especially obtuse topological formula, I did some chores outside in brisk weather – the presence of the *whole body* being what's typically missing from the textbooks we get. I was cleaning the dust out the large tank I use to collect rainwater by adding a little spare water and sloshing it around, trying to control the flow of particles around the edges of the drain valve. That same night, I had a half-waking dream of dust suspended, gently levitated and settling down upon a curved surface: I knew that the particles were the underlying set, their distribution upon the surface was the topology on that set.

- 124 -

> Most of the information conveyed by a sensory neural spike train comes from the origin of the signal. Every patch of your skin contains a variety of specialized sensing neurons, each conveying a specific message. Similarly, the result of motor control signaling is largely determined by which muscle fibers are targeted.
>
> J. Feldman, "Ecological expected utility and the mythical neural code"

The idea we're exploring is that the topological space of neuronal information is layered on top of the affine structure of proprioception: thus the tension between the distorted cortical homunculus and the precision of the cerebellar calculus. Thus also, many of the more notorious hallucinations we get via psychedelics and sensory deprivation: the systems are allowed to decouple, and we begin to see behind the curtain of the elaborate opera called neuromorphic experience. What is "raw" neuronal information but the excitation patterns of a collection of overlapping subsets, which has no inherent "location" except through their indexed values in some other receiving system? If we could ever get close to this this raw state, wouldn't it seem as though our body were floating, or the homunculus proportions made our hands and face seem huge, or that our skin were translucent, or that the essence of the body were a fuzzy ball of light...

Neuromorphic experience begins as a topological space, which in the case of proprioception is reconciled to an affine structure, which then is enacted in the real Euclidean-like world, and which receives feedback therefore from that same 4-dimensional world. The fate of a proprioceptive neural bit: radical informative reduction to excitation patterns of overlapping sets of neighboring neurons, which is indexed against a cerebellar structure, which is then selectively reprojected as needed for precise movement, and which implicitly establishes feedback channels through afferent-efferent loops in tissue and the overlapping sensory systems. Again we must remember that information is not information until it informs *behavior*: the neuron was, is, and will continue to be a muscle actuator exploited for its capacity to carry feedback the other way... The body is a circle.

- 125 -

There is no metric in biological systems. No dimensional axis, no coordinate system, no Pythagorean distance function. There is *ratio* of lengths, and there is continuity between overlapping sets. An organism measures the world with its own body.

- 126 -

Neuronal information is minimal: it is metabolically optimized. It is in the *indexing set* of the receiving systems that the information gains back its richness: a point in space can be estimated with high accuracy not because of the specific accuracy of any one sensory neuron, but because of the high degree of overlap between multiple simultaneously indexed sensory inputs. Navigating a neural topology is like picking one's way through the web of any healthy ecology, or the interval structure of a rich melody which implies many excursions it need not take.

- 127 -

We study topology because it yields insight into the mystery of proprioception, and proprioception is the *fundamental sense*. Every "feeling" depends upon a *localization function*: the ancient and worldwide tradition of locating the source of emotional information in the interior organs – the *heart* and *intestines* above all – is not because our ancestors were stupid and superstitious. It's an adaptation of the vertebrate nervous system, to perceive emotional valence as a kind of inward extension of the proprioceptive function: granted that emotion itself is defined as preemptive adjustment to behavioral threshold – the animal must feel *compelled by inner necessity* in the way it is compelled by outward obstacles. Emotional intelligence is therefore a kind of efficient exploitation of the *capacity for navigation*: we pick our way through a feeling. Thus is topology valuable, because this is a *place without place*, an imaginary space without metric – but one in which the preserved properties of space hold, such as continuity. "Can I get there from here?" : that's the question every emotional navigation asks. Such is what we develop both in meditative practice and good psychoanalysis: the navigation of continuous and discontinuous undulating sets, treacherous islands, molten footholds, hidden precipices and invisible bridges.

- 128 -

Yet in practice, what makes the difference between growth and stasis? A spiritual life depends more than anything on the ability to accept *not knowing* where you are, nor how you'll get to where you want to be. It demands navigation of the unknown with minimal information: everything else is repetition compulsion and the monotony of neurosis. A neurotic knows where he is, because he knows how to distort every novel uncertainty into monotonous certainty: he's always going the same way to the same place with the same means, no matter how diverse the context. When we meditate, the challenge usually distills to this: overcoming the thousand-and-one means of escape from the omnipresent unfelt feeling, the ever-louder unthought thought, the unacceptable unconscious conviction. It's not that people don't know how to navigate complex topologies – on the contrary, masterly unconscious distortions, dizzying twists, clever reversals, folds within folds, and denials hiding assertions hiding denials, are the norm in the human race. *Staying put*, gently feeling the edges of the basin, tracing lines of descent and ascent, accounting for one's twisted limbs – in short *waking up* to the proprioceptive reality no matter how steep the incline and precarious the position, is precisely why so much courage is required. The sicker we are, the more harsh this awakening – which is why the earlier we begin, the better. We all have had that long backlog of overdue pain tumble to the floor with a crash... Psychosomatic disease is a waiting room.

NONLINEARITY

The straight line is a curve, any part of which is similar to the whole.

Leibniz

\- 129 -

What is nonlinearity? In its simplest sense, it just means *curved*: when one variable responds exponentially to another. There is almost nothing of interest in this world which is linear. Linearity is only ever approximate, and strictly speaking only exists in ideal models where friction, perturbation, and resonance have been subtracted. The bias towards linearity in contemporary science is an artifact of the way Greek and Babylonian mathematical approaches were abandoned in favor of purely algebraic reasoning as developed in the Islamosphere and late-Medieval Europe: linearity suited the fundamental assumptions of the postaxial world, and was not seriously challenged even when Newton and Leibniz developed calculus from those assumptions. Calculus is still taught in a rather clumsy way which assumes linearity as the basis of the study of change, and it is only in trigonometry that the *circle* and the *ratio* is granted prime citizenship again, as it was among the ancients. Remarkably this ancient mode survived in its stronghold of astronomy almost until this same point, exemplified by Kepler's solar system model: his dominantly *geometric* reasoning is evident – he thinks in shapes, rotations, angles, ratios.

\- 130 -

Nonlinearity is everywhere, and yet contemporary science is replete with obtuse linear assumption: it wants linear formulas for as long as it can get away with it. Nowhere is this more egregious than in neuroscience and cognitive psychology.

- 131 -

Again what is nonlinearity? In systems theory, it is *the effect of history* – what we call "feedback" is nothing but the effect of history: a linear equation may express the resulting feedback given some input, but not the *accumulation* of feedback effects given some history of input and response. A nonlinear system *has a history*, and this history expresses itself through the responses of too many parameters to neglect: there's rarely a "closed-form solution" to a nonlinear dynamic system, despite being fully deterministic. Yet feedback exists in almost every context of interest, because almost every context can be construed as a bounded system wherein previous states alter its subsequent responses.

- 132 -

What is a strange attractor? It is the result of a concave statespace. Consider the pebble: why is a pebble round? A pebble is round because a polyhedral rock under the influence of water, tends to roll. The more it rolls, the rounder it becomes, and the more it rolls. Thus a spherical pebble is an attractor in the bounded system of a rocky beach. It is that simple, that ubiquitous, that profound.

- 133 -

"Nonlinear" often means *not a summation of its parts*, and yet characterized by identifiable behavior. The responses of nonlinear systems produce *shapes*: shapes which hint at the internal dynamics of that system, but which themselves do not yield to simple formulation. The attempt to express the behavior of a complex nonlinear system in terms of a bijective function, tends to result in a model with as many moving parts as the system itself. Thus the best chance at an elegant reduction is to map its statespace and look for patterns and phase transitions. In a sense, a constrained complex nonlinear system gives away its secrets at its critical moments: as though the underlying dynamics reach full expression at the limit of their functional capacity. Before and after that point, intervening dissipative modes and the coexistence of multiple order parameters tends to blur the potential for emergent behavior: only when a system is forced through external driving to its critical moment, does one set of those many possible dynamics dominate to the degree that it yields to measurement and formula.

- 134 -

The simplest way to explain nonlinearity to a musician, is via the concept of *resonance*. When you strum a guitar, what you hear is not only the sound of the string alone, producing sine waves as though it were suspended in space, but the sound of that string's vibrations filtered through the body of the guitar and their interactions with each other within that body. This body has a nonlinear response: not only do the resulting frequencies fail to map in a one-to-one relation to the generative frequencies of the string, but it is neither exponential nor logarithmic nor anything else easy to model. The output response is unique to every guitar and *expressive of the shape* of the instrument itself, something which is not surprising since in this case the output response is nothing other than a pressure wave in air generated by a resonating rigid body. If you were to plot the frequency response of any given instrument, it would create a unique shape, the signature of that instrument, its *timbre*. Thus it is that musicians speak of one sound as "warm", the other "harsh", "tinny", "muddy" and so on: they're describing nonlinear response graphs at various amplitudes. This is part of why for example, modern computer generated music relies on *sampling*, rather than pure synthesis, since it is so difficult to synthesize from a sine wave the warmth and complexity of a human voice or a cello.

- 135 -

Meditation is largely the exploitation of those same nonlinear dynamics, stochastic resonances, and feedback channels which the vertebrate nervous system was originally constructed around, to achieve *neurochemical modulation*. When we meditate, we are tweaking the parameters of a self-regulating nonlinear network topology, which *necessarily* responds to such tweaking because self-regulation is essential to its adaptive function: it's not magic nor transcendent nor holier-than-thou, except to the degree that the vertebrate nervous system is all of those things and more.

It's possible for the human creature to use apperceptive consciousness to modify its own neurochemical homeostatic function: that's the miraculous, overused, underappreciated power of apperception.

NONLINEARITY

OUR PHANTOM LIMBS

- 136 -

When Oliver Sacks lost half his vision to eye cancer, he characteristically put the condition to use in research: he began experimenting with this blind spot and watching the resulting hallucinations. He found that he could virtually amputate his hand:

> But if, when I had done this, I opened and closed my fist or moved my now-invisible fingers, a sort of pink protoplasmic extension grew out of my visual "stump" and developed into a (visual) phantom of the hand.
>
> *Hallucinations*, p.284

Something like 95% of all amputees experience phantom limb syndrome – it's even relatively common with congenitally missing limbs. This is another example of a commonplace mystery which is greatly occluded by commonplace prejudice: it required an honest doctor exposed to the fallout of the American Civil War before this syndrome was even discussed openly. But the story goes much deeper than amputation, and involves every healthy limb: the phantom limb is not merely a phantom of what was lost, but the only limb we ever experience.

- 137 -

> [...] the completely static picture of these maps that you get from looking at textbook diagrams is highly misleading and we need to rethink the meaning of brain maps completely. Far from signaling a specific location on the skin, each neuron in the map is in a state of dynamic equilibrium with other adjacent neurons; its significance depends strongly on what other neurons in the vicinity are doing (or not doing).

V.S. Ramachandran, *Phantoms in the Brain*, p.35

Amputees often complain that their phantom is contorted and clenched, and the pain is excruciating. The reason seems to be that without calibrating feedback, the proprioceptive projection will persist without attenuation, as though the threshold of perception were continually lowered to its minimum in order to discover the missing afferent input. The pain seems to be a result of the destabilization between attenuating feedback and a variable threshold: most perception is a *perception of a projection*. But what's even more surprising, is that this kind of pain can be managed by taking the phantom seriously: many amputees feel that their phantom is terribly restless, but the addition of a prosthetic calms it – the "lost phantom longs for a new home", as Sacks says.

The longer a neurologist works with phantom limb syndrome, the more the doctor's professional arrogance is eroded, and such statements begin to seem like the only humane and sensible approach to the problem: a phantom limb is no mere illusion, no mere error, but the living envelope within which the flesh is allowed to move. It is the command matrix of the body, generated continuously – as though the muscular obedience were almost an afterthought, an inevitability.

In the 1990s, the neurologist V.S. Ramachandran developed a powerfully simple therapy for phantom limb pain: the patient places his amputated limb behind a mirror, and arranges his existing limb such that the mirror generates the illusion that the amputated limb has been restored. With this single trompe-l'oeil, the patient can often unclench a phantom hand and relax a phantom leg. Such is the power of apperception.

- 138 -

There's sufficient evidence that significant cortical remapping takes place after amputation: the cortical homunculus shifts, often collapsing the hand representation into the face area. However, the phantom limb persists: the slowly growing consensus seems to be that the so-called "motor command representation" is produced otherwise and elsewhere, and is the basis of this phantom. But what very few researchers seem to have internalized is the importance of the cerebellum, despite everything we know about the consequences of its damage: ataxia, vertigo, imbalance, a wobbly gait, nystagmus, difficulty targeting and correcting both finer and broader movements of the body. What's rarely mentioned is that *80% of our cranial neurons are cerebellar* because it's such a dense and efficient calculator. I'm afraid that the neglect of the cerebellum in neuroscience is once again rooted in some pathetic reasons of ape vanity: it's one of the oldest structures of the vertebrate nervous system and ours is nothing special. It would be difficult to discern any functional difference between the human and reptile cerebellum. All the more reason why it should be considered as the most likely source of coherent and precise movement, since this is something our fish ancestors mastered a long time ago.

- 139 -

What's the function of apperceptive consciousness? That distortion and amplification is a normal side effect of trance should give us the clue. That feeling of holding the heavy ball, of the succubus, of the homunculus emerging, "my hands felt just like two balloons": aren't these experiences witness to the tension between the nerve density mapping of the cortex and the spatial resolution of the cerebellum?

So much of "hallucination" is merely exploration of lateral routines further up the chain of processes, which normally remain inaccessible or hidden within the result. Back a step, and to the side: inhibition of spatial resolution, and visualization of the afferent network using an equidistant metric produces this "sensible homunculus".

Original greyscale

Monochrome threshold

Monochrome dither

Oliver Sacks circa 1961.

- 140 -

It's not enough to say: "this afferent data should be reorganized like so." It must actually put the system through a diagnostic routine: it must excite both afferent and efferent centers, so that the two cooperate.

Excite afference and efference: this means, lower firing thresholds. Possibly resulting in lowered nociception thresholds, which accounts for so-called "pain wind-up" and the psychosomatics of both phantom limb and effective massage. How is pain "stored" in an area? Is this pain the result of dramatically lowered nociception threshold at the moment of proprioceptive reorganization? In good massage, the danger is "tensing up" as the musculature shifts while proprioceptive remapping takes place. Releasing tension and proprioceptive remapping are *always coincident*: the body thought it was "supposed" to be in that tense position.

- 141 -

During Ramachandran's phantom limb therapy, it's common for the amputated limb to twitch and tremble. Even just talking about the phantom can induce it. Neural noise is added, an excitation that resembles the *anticipation* of sensation – a prickling, burning, tickling. This localized excitation is not just an artifact of amputation but an artifact of the latent apperceptive loop: neural noise is added in anticipation of perception, and seems to increase in the presence of error or delay. Why? The answer lies in something called "stochastic resonance": under certain circumstances, the addition of noise creates a clearer signal.

Stochastic resonance is why we jiggle a lock and used to smack old TVs. By boosting amplitude across all frequencies, latent signals may have a better chance of surpassing the threshold of activation – information theory doesn't care if the signal is a doorbolt or a neural spike train. In signal processing, this is called "dither", seemingly originating in WWII bombsights:

> Airplane bombers used mechanical computers to perform navigation and bomb trajectory calculations. Curiously, these computers (boxes filled with hundreds of gears and cogs) performed more accurately when flying on board the aircraft, and less well on ground. Engineers realized that the vibration from the aircraft reduced the error from sticky moving parts. Instead of moving in short jerks, they moved more continuously. Small vibrating motors were built into the computers, and their vibration was called dither from the Middle English verb "didderen," meaning "to tremble."
>
> Ken Pohlmann, *The Principles of Digital Audio*, §3.2

- 142 -

In acoustic terms for musicians and audiophiles:

1. Nonlinear response to suprathreshold amplitude creates a sparse signal with harmonic distortion. *Harsh*.

2. The addition of noise blurs the harmonic additives, by probabilistically raising all frequencies, thus obtaining a flatter distribution. *Muddy*.

3. A lowpass filter attenuates the whitenoise which was pushed above threshold. *Warm*.

The result is something like running a hollowbody guitar through a vintage vacuum tube amplifier, which has that characteristic warmth so difficult to reproduce.

- 143 -

> [...] the parameter configurations that set the system to a critical state coincide with the parameter values at which the transmission of a signal through the network is enhanced.
>
> Vázquez-Rodríguez et al, "Stochastic resonance at criticality in a network model of the human cortex"

The optimum noise complement for stochastic resonance to occur in any given nonlinear system seems to converge to a critical point of phase transition in that same system. But that would make sense if the relevant signal gain is conceived as a kind of exploitation of phase transition: amplitude above phase threshold produces a nonproportional response, which effectively amplifies an otherwise weak signal.

- 144 -

Given that vertebrate hearing resolves both pitch and amplitude along a logarithmic scale, is it possible that every perceptual field is similarly arranged to attenuate higher frequencies, in order to approximate a `1/f` inverse power distribution – also known as "pink noise"? If that's true for proprioception, is part of what apperceptive consciousness does is to add back a higher frequency noise complement – something approximating white noise – in order to achieve stochastic resonance?

We've been playing with at least two models accounting for resonant neural feedback: topological tangential space sampling, a kind of projection into a low dimensional plane; the other is the addition of noise to achieve stochastic resonance. Can the two be reconciled?

The hypothesis is that the relevant features for sampling are selected through a gain in static in any given neuronal system, such that salient features percolate out and are then reprojected and fed back into the system.

1. Stage one excites the neural network via the addition of noise, which thereby obtains a rational sampling.

2. Stage two routes the clarified and coherent aspects of the topology thus outlined, to an apperceptive projective space which we imagine lies in the sensorimotor cortical network, which is then fed back to a higher-resolution unconscious space lying probably in the cerebellum, thus amplifying the most salient features of the signal. Conscious apperception is effectively surjective due to the forced resolution and lowpass filtering of outliers in the process of producing a sparse and projected topological image. Therefore the purpose of apperceptive consciousness is to produce a *resonant skeleton* of a perceptual network in order to improve signal transduction, response to relevance, and an optimized metabolic flow.

- 145 -

That meditation is the practice of *localized criticality*: the application of both periodic forcing via topological projection and an increase in additive noise, such that what we are aware of becomes *more likely to spontaneously activate*.

The more we practice this, the more our tolerance for ambivalent excitation grows, and the deeper our intuition becomes concerning where the "sweet spot" of any given activity lies: at the transitional moments, between normalized modes, such that a minimum energetic expenditure is required to achieve maximal bursting – we become funnier, more likely to laugh, more likely to feel novelty...

- 146 -

The study of phantom limb syndrome is not entirely theoretical for me. When I was 17, I was working in construction and sustained an injury which has turned out to be full of fateful implication: my left thigh was cut open with a circular saw down to about one inch, leaving a scar about 8 inches long. It was nearly fatal, since the blade ran very close to the femoral artery. The muscle healed well enough, but the nerves did not: there is still a large numb region below the scar. The remarkable thing is that a small region around the knee seems to innervate both the knee and the thigh: I feel sensation in both areas when the knee is touched just so. But the resulting sensation in the afferent dead zone of the thigh is confused, chaotic, sporadic, and often painful. It's somewhere between tingling, burning, and an *anticipation of sensation*. Moreover, the injury left its mark on my proprioceptive network: there is a kind of hole, which pulls at the rest of the affine structure below and above, twisting it and distorting the image. It's caused referred pain, and subtle misalignment which I have revised for many years through Qi Gong, yoga, and meditation. In fact, repairing this damage and the resulting psychosomatic pathological cluster, has been my secret weapon in this study: disease is the great teacher. Error is the best microscope.

When I read about phantom limb syndrome, this part of my left thigh twitches: which means that merely the apt description of the syndrome can induce it, something which any good doctor has long known – and only another entry in the giant neglected file called "psychosomatism". I believe that what I have is a small lacuna in the affine structure of proprioception, which causes a distorting tension to propagate through the rest of the network: thus referred pain, thus changes to gait and stance which cannot be explained in purely mechanical terms.

This wound has taught me: I would never have learned to refine meditative practice to what I believe is its essential power, if I had not been forced by pain. Apperceptive consciousness conjoined with breath control can permanently alter proprioception, nociception, and the entire affine structure of the "bodyimage": but the breakthrough came when I realized that this kind of modulation and diagnosis is precisely the function of apperception – consciousness is a diagnostic routine, which should properly leave behind a more orderly and efficiently traversed graph than it found...

- 147 -

What is it like, when I heal this lacuna? It is a prime example of the cooperation of respiratory and apperceptive modulation. In the language of modern yoga, I "send breath" to that region: what I often visualize is that breath knits a temporary web which pulls the network into place. In more technical terms, I suspect that apperceptive consciousness is locally raising the ambient noise complement, while the respiratory centers responsible for exhalation in the medulla oblongata amplify parasympathetic response, which induces the neighboring regions of parietal lobe associated with the relevant portion of the proprioceptive map to communicate with each other and adjust to the informative gap. The practiced diaphragmatic exhalation induces a reparative routine, and prevents the interference of amygdalic or sympathetic activation, such that the anticipation of pain is inhibited, even while chaotic information arrives at the cortex which would normally signify a dangerous degree of injury.

When done properly, what often follows is a kind of *myoclonic cascade*: a localized spasm which releases accumulated tension in the affected neural systems. I have learned to exploit this side-effect to its maximum benefit: I'm able to induce a tightly controlled "microseizure", the aftereffects of which leave me feeling refreshed – much like migraineurs feel the next day. But the story goes deeper still: what I've discovered is that once a major locus of trauma exists in the nervous system, it exerts a gravitational effect on all other forms of psychosomatism. An asteroid belt of woe, pity, and self-destruction is likely to form: especially in we delicately balanced modern types, who have already undergone so many forms of subtle abuse in the midst of punitive normalcy – something's gotta give. And it is only we fortunate ones who end up with interesting and sublimated syndromes, like migraines and esoteric boutique microepilepsy. What I can tell you, is that pain seems to enter and come out of that hole: if that sounds like a hysterical formation I cannot deny it. But what everyone seems to forget is that the essence of hysteria is *displacement*, and the reasoning here becomes involute: displacement occurs because of *repression*, which is effectively a "dead zone" in symbolic terms. Displacement is therefore a kind of network distortion as a result of this hole. It is one of the prizes of a psychoanalytic education to be able to detect distortion in the so-called "semantic field": therefore the fact that measurable physiological damage to a proprioceptive network might mirror the effects of the so-called *purely psychogenic*, should not surprise us. But the relation is not merely one of parallelism: everything psychological is an emergent property of the physiological. There is no distinct frontier. And it is not poetic sloppiness, but part of our graduated perspective to employ carefully selected psychological vocabulary in the realm of physiology and physiological terminology in the domain of psychology. This numb region on the surface of the skin is a kind of "repressed" zone, and around it is a distortion field of displacement. This makes the application of topological terminology to psychology seem all the more relevant and piqué.

- 148 -

What I'm implying is that phantom limb syndrome reveals the *template of hysteria*: that possibly every extension of the proprioceptive field has a protohysterical component. That *Besetzung* is a species of *Verschiebung*: that our tightly controlled extensions in meditative practice as an exploitation of tool use, is therefore a kind of "reflexive cathexis".

\- 149 -

Components of the theory:

1. The function of the cerebellum is a *spacetime calculator*: extremely fine calculations of distance, angle, and momentum.

2. Proprioception is realized as affine structure, primarily generated by the cerebellum but adjusted by visual and motor centers on the cortex.

3. Apperception is a localized calibration routine of this collective proprioceptive network.

4. Apperception functions by:

> ⌐ *Inducing stochastic resonance* to gain better mapping resolution, and possibly evoke emergent structure. Evidence: tinnitus, pseudoepileptic symptoms in psychedelia, serotonin overdose, and deep meditation.
>
> ⌐ *Inducing affine structure recalculation*, via graph traversal. This is possibly also a consequence of stochastic resonance excitation. Evidence: induced twitching during phantom limb therapy.
>
> ⌐ *Inhibition of cerebellar spatial resolution*: integrating feedback from the cerebral cortex, especially visuospatial results. "Seeing" what we feel is a side-effect of the integration of these high-performance processors. Evidence: the role of visual input in the "mirror box" treatment for phantom limb syndrome. This kind of apperception may be dependent on the inhibition of finer spatial resolution, such that afferent densities and weights become sensible, or even adjustable. Evidence: the hallucination of the liberated homunculus. This inhibition of spatial mapping and access to afferent densities, may be an artifact of visual cortex involvement: it begins to visualize the densities, without reference to the

spatial displacements... An *equidistant metric function* between afferent nerves produces the sensible homunculus.

- 150 -

One of the most important lessons I learned in my days as an engineer, is the importance of *scalability*: it seems to me that very few biologists have internalized this principle, nor learned how to apply it when theorizing about living systems – however, there are signs this is changing in the field of genetics. For example, the accumulating evidence for the ubiquity of *facilitated variation* in the inherited genome: ontogenesis doesn't code for precise locations for nerves, veins and other networks in the body, but runs precisely tuned algorithms for the *process* of innervation, vascularization, and so on, which respond to the growing bones such that only a few parameters need be changed to produce either a mastiff or a chihuahua.

So what makes for scalability?

- Minimal dependencies
- Minimal representation
- Rigorous specification of homogeneous units
- Redundancy
- Error recovery

The more that a field of interest can be characterized as a collection of interchangeable units with predictable properties, the more scalable it becomes. This also enables *redundancy*, since many such objects can perform overlapping tasks – such as many nerves serving the same region; thereafter *Byzantine fault recovery* becomes easier due to redundancy.

It may seem like vulnerable design to form a neural topology dependent upon its constituents in absence of a global coordinate, but the probable truth is that errors generated by any one subset are generally compensated for by its neighbors – and that the kind of error we've been considering in phantom limb syndrome, only occurs when a significant portion of the network is missing.

- 151 -

The prevalence of facilitated variation in everything the organism does, should make enduring neuroplasticity and the *regenerative character of apperception* as I'm portraying it, seem a little less exceptional. Neural systems are just like everything else organic: they must be able to adapt to a changing environment, and not just during ontogenesis. The reason meditative practice can have such enduring effects, is that despite the "hardwiring" of dendritic architecture, it seems that most neural activity is a function with many parameters and few constants. This is what connectionist artificial neural networks reproduce, albeit crudely. It seems improbable that *backpropagation* adequately reproduces neural function: it's the rare situation in which a precise answer to "how correct was the last guess" is given to the organism – another artifact of the bad assumptions of the laboratory clerk. However, feedback is obviously involved: so much so, that models of self-organizing criticality and stochastic resonance are much more convincing, and reminiscent of what it feels like to be alive.

- 152 -

The idea I'm trying to get across is that there are at least two primary spacetime and body mapping functions at work in the vertebrate neural system: the more permanent topological characterizations maintained by the cortex, and the affine calculations generated ad hoc by the cerebellum. The trick of Ramachandran's mirror box therefore, is to force the more static yet consciously accessible cortical mapping to adjust to transient cerebellar estimates. A *mirror* is needed because an overrepresented topology handles gaps poorly: neurons within and near to the gap, now lacking inhibition from their silent neighbors, fire in an overactive way which simulates pain and prevents remapping. This remapping scenario is possible, but it requires not only stochastic resonance to produce a newly coherent image, but the quiescence of other functions in order to acquire the neural bandwidth for so expensive an operation. This theory accounts not only for the symptomatic *twitching* which accompanies phantom limb therapy, but the need for parasympathetic activation, namely *calm slow breathing*. This profile of a successful phantom limb healing session, also coincides with excellent meditation practice: we unlearn so that we may learn. We get excited to calm down.

EMERGENT COHERENCE AT CRITICALITY

- 153 -

What we achieve in meditation is not hallucination in the usual sense, but rather these "alternative states of consciousness" are byproducts of *emergent coherence at criticality*: this coherence does not exist outside the immediate functioning of each respective neural system, and is therefore "subjective", ephemeral, and difficult to formulate. The meditative art yields to statespace mapping and power-law estimation, rather than reductive analytics. It is a nonlinear miracle of a vast complexity, and therefore is not "in" the physiology outside of its critical moments – subjective experience cannot be discovered via vivisection: this is why it has not yielded to the typical brutality of 20th century medicine, which generally believes in nothing until it's splayed out on the laboratory table. But there are signs this is changing: neuroscience has been creeping up on nonlinearities with increasing assurance, and it's time that we spiritual practitioners caught up.

- 154 -

Modulation of the nervous system would be a relatively trivial exercise, and perhaps just as self-indulgent as the detractors of meditation might wish, if the sole function of the neural network was to generate immediate experience – as though that were an end in itself! The body would not perpetuate such lavish energetic expenditures without good reason. Subjectivity, affect, a sense of being-in-the-world, has discernible and exploitable function just like everything else in biological systems: the nervous system *coordinates*, *regulates*, and *navigates*. Therefore when we meditate, we are adjusting and refining the primary short-term regulatory system of the body: we use selective feedback to modulate this "subjectivity generating" feedback system which modulates many other systems in turn. This is why meditative discipline has the power to change your life entire. Keep in mind that we are much more than merely the epiphenomena of the nervous system – to say nothing of modernity's habit of placing all personhood solely in the brain, and sometimes only the cortex.

Meditation is therefore the exploitation of the side-effects of extremely short-term emergent coherence within a system which itself has a short-term regulatory role situated within longer-term systems – such as the digestive, the immune, the hormonal, or the skeletal-muscular systems. Every time we successfully meditate, we leave traces which latent neural function becomes more likely to follow: often it is only a few seconds of novel configuration within a 30 minute session which represents our progress – but due to enduring neuroplasticity, it is enough. This is why meditative practice deepens and evolves over the years, as we eventually need it less and yet also learn to exploit it more.

The regulatory leverage of the nervous system also helps explain why we seem to gain in overall health through steady practice: our digestion becomes more regular, our immune system kicks ass, we sleep better, et cetera. Much of what it means to be healthy amounts to rapid and distinct transitions between activation and rest: nothing characterizes the chronic illness of the modern body like endless low-level inflammation – always half in, half out. Always lazy, never restful. Always wary, never awake. Always craving, never hungry. Always irritated, never enraged. Always smirking, never joyful.

- 155 -

I frequently use the word "coherence" when speaking of meditation: why? In signal analysis, *coherence* means rational waveform, constructive interference, and implies a musical relationship. But the brainwaves measurable via EEG are most likely a kind of baseline coordination channel, sufficient for low density information and general inhibition – "hey I'm still here": the days when we thought that such waves could be a sufficient carrier for global coordination are over. This is why neural criticality has become important, because it seems to be the approximation of fractal geometry which yields poised susceptibility and the avalanche: in other words it's not bruteforce phase synchrony but *transient associations* producing the computational power of a bundle of neurons, but we still don't know how these associations arise nor how they encode long-range signal.

What we do know, is that so much of what's wrong with us is a result of incoherent interference between systems which should properly nest one inside the other, overlap, and mutually reinforce – resonance is a symptom of health. Microscale coherences such as we create in meditation have the power to engender macroscale systemic coherence – just as for example, rigorous daily routine and regularity of diet are part of the recipe of maximum human development.

What are these "microscale coherences" I'm speaking of? Partly they are the recruitment of artifacts: in other words, the almost incessant liminal hallucinations and probable errors normally inhibited, blurred, and "lowpass filtered". Because of the stillness and deprivation of meditative practice, these processing artifacts surface in a way vaguely analogous to what happens in psychedelic experience through more forceful chemical means. *Tinnitus*, for example, is both the result of a sudden loss of input and too high a gain in the noise component added to achieve stochastic resonance in the elaborate resolution functions of our hearing. It may go by many other names in the history of spirituality, but the neural noise I'm speaking of, which when constrained and thus amplified, is a kind of "apperceptive neural tinnitus": apperceptive consciousness in combination with the modulation of breath control has the capacity to locally excite this latent stochastic static and force a subset of the neural network to undergo a phase shift to some other stable limit cycle.

In the past, this latent stochastic noise and the control of its emergent properties would have been called "qi", or "kundalini". As far as we can tell, most of these emergent orders are extremely transitory and difficult to reproduce: however we all have our accustomed tricks and preferential means of trance.

So if we grant that healthy neural systems operate near a critical point, and if we further grant that meditative discipline induces and refines the exploitation of this arrangement, which must include long-range correlation and an approximated scale invariance in activation patterns, then it follows that part of the practice must be the tolerance and amplification of slightly strange peripheral noise, which becomes more sensible and rational over time. It's telling to me that despite all the attention which self-organized criticality has received in neuroscience, very few have ever asked themselves what it might feel like to undergo a near-fractal activation – whether it might be sensible, whether it might be overwhelming, whether fear or rapture might be involved.

The point is that we are not merely playing with ourselves and inducing useless hallucinations: forcing the system to proceed through successive modes such that it effectively explores its own harmonics, not only leaves behind traces in the system such that it becomes more likely to visit these stabilities in the future, but it also tends to induce *novel reparative subroutines* dependent on these higher dissipative states. Parallels in complex biological systems are not difficult to find: such as what happens when you force a heavy smoker to undergo a strenuous hike until they vomit, and find a surprising willingness to keep going. "Second wind" is another way of saying "dependent subroutine", and there are many such modes of the stronger body latent and waiting: most of us do not know half of what our body can do. Much of what occurs in the early stage of meditation is the analog of *puking your guts out*: I know a woman who during the first year of her practice, would take five minutes to reach a certain meditative state and then spend the next hour weeping. She would exit this practice clean as a whistle and supremely receptive.

- 156 -

Actually hallucinatory control and self-induced trance is not at all unusual: even the most sedated and addicted modern human body glides without much trouble between various forms of trance. Everyone you know is adept at producing *at least one kind of trance*: the difference is the monotony and compulsion involved. But in the language of neurophysics, what is *trance*? It is sustained inhibition of an otherwise dominant mode or affect cycle, such that latent responses have a chance to develop and reach expression. This is analogous to the tranquility needed in the confabulation cycle needed to produce novel ideation: in a healthy system it is the moment of eureka, epiphany – stopping mid-sentence and staring into the distance with a gleaming eye. However this kind of broadband inhibition tends to also be the signature of the coping entrancements of modernity: this is why everyone seems like a zombie, shut down, half dead. It is also the reason why the masterful hypnotist Milton Erickson, for example, generally found it necessary to jolt the subject awake first, in order to induce some other more productive trance: therefore meditation is not "zoning out", not "checking out", *not essentially escapist* – although I'm well aware that it's been abused as such, and even *advertised* as such for millennia. Meditation in our hands becomes a weapon *against escapist monotonous trance*: we learn to proceed through our phase shifts ever more fluidly, ever more rapidly, we learn to feel our feelings more deeply, more passionately, more committedly. *The grand cumulative trance of being alive*: that's the bigtop under which all these exercises should spin. We are more sober than anyone, and simultaneously more intoxicated than anyone with the elixirs of life. We're raucously drunk at the fourway marriage of O, C, N, and H.

- 157 -

Hypnosis is generally achieved as the inhibition of executive centers through the medium of social authority. Most of humanity is so suggestible because it is socially desperate. In addition, the more alienation and an impending sense of abandonment looms, the more willing the tribal creature becomes to suspend executive function for the sake of inclusion at any cost. Those familiar with my previous work will know this. But those destined for the most profound meditative practice, are usually immune to such means of hypnosis: we don't meditate because the Buddha wants us to, or because the sangha is watching, or because we expect to find an approving godhead at the end of that tunnel – or any other form of moral prostration.

Yet we induce trance in ourselves: how? Partly through this same function of authority. It's possible that what's meant by "willpower" is nothing other than this same tribal authoritarian hypnosis applied in a reflexive mode: we obey ourselves. Generally the most willful human creatures are not happy until they are dominated by something tyrannical: some nearly impossible task, some great ambition, perhaps a lost cause, or a great love. And while setting the intention to meditate is extremely important, and investigating the means of establishing a challenge-and-reward schema relative to the practice is vital – you have to *want* to do it – this is rather the *longterm* profile of a meditative practice. Something else occurs in the immediate inducement of trance. I'll skip to the end give away the answer: it's about the ability *to say no*. Trance is conditioned by *massive systemic inhibition*. Therefore the cultivation of attitudes of withholding, rejection, and a slow affect is vital to the development of a meditative practice. The wrathful deities have been considered useful because *wrath is useful* for the generation of containment and neural inhibition necessary to induce deep trance. It's not just so that we can overcome anger, but to exploit it. Everything has a place at our table, because we seek to be complete human creatures, and because our discipline is beyond good and evil.

THE RUBE GOLDBERG MACHINE OF MEDITATION

- 158 -

It has been dawning slowly, achingly slowly over the last thirty years upon the more prepared minds in biology, that nonlinear emergence may not only be peripherally involved in the optimization of biological function, but lie right at the heart of the definition of life. Look at it from the purist Darwinian perspective: life is exploitation – exploitation of energetic surplus, exploitation of every niche, exploitation of every strategic position. If therefore any advantage were to be gained in a complex dissipative system from feedback, resonance, and self-organizing criticality, then surely life at its very roots would have already exploited these advantages. Indeed it's possible the statement should be reversed: that the key moment was the *recursion* of self-organizing criticality in the emergence of the *genetic* facility. In other words the ability to propagate an order parameter while retaining *dormant traits*: traits optimized not just for one phase but also for some other critical phase - and thus a "lifecycle" of this emergent organism is born. Self-organization is an essential but insufficient sign of life: an avalanche is not yet alive, because while it may exhibit scale invariance and self-organization such that it has "offspring", it does not retain traits and thus cannot adapt. A *dormant order parameter* is required: a phase transition within molecular noise which produces a recursive critical moment within its own competing varieties of self-organization over multiple phases. Genetics too, may be constituted by something like stochastic resonance...

I am not at all the first to intuit this. You could trace a genealogy back through Hermann Haken, Erwin Schrödinger, Nietzsche,

Goethe, Spinoza, Aristotle, and Heraclitus. We are nowhere near the final revelation: more likely we have only begun to peek behind the veil. But for our purposes this can be safely emphasized: feedback and modulation are no peripheral accident in the formula of optimal health. Therefore whatever means we can find of tuning these modulatory functions, should be handled carefully and in general be given a great deal more attention and respect than hitherto. My contention is that not only is the respiratory center in the brainstem one of the principal modulatory powers available to us, but *consciousness itself*: my theory is that we do not begin to understand consciousness until we ask what its *adaptive function* could be. For too long, everyone has assumed that self-awareness is a self-evident end-in-itself. There are at least three snakes with tails in their mouths in that last sentence: the presence of that much *contradictio in adjecto* should give us pause. There is no "in-itself", nor is there "end-in-itself", while "self-evident" is an idiomatic hyperbole, and so is perhaps "self-aware". Perception always implies something which *cannot be perceived* – Kant taught me at least that much: perspective implies hidden origin. There is awareness, but there is no "aware self-awareness" such as the few *consistent* and persistent philosophers would like to discuss: this kind of scholastic nonsense already peaked in the likes of Schelling, Hegel, and other Upanishadic wannabes – but at least these types had the decency to draw articulate conclusions where everyone else in our time is content to *imply and assume*. My kind of reasoning leads elsewhere: *there is no "conscious self" to be conscious of*. But the next step is not metaphysical despair at the loss of something we didn't need nor ever possess: these are all shallow, relatively recent postaxial priestly snares, delusions, and consolations. The Buddhists are correct about "anatman", but they are utterly wrong to consistently portray this as some great loss which we can expect to be recovering from for the rest of our lives. Nothing is lost but a little stupidity, a little rigidity, a little unwillingness to learn. Our biological inheritance is *functional* above all, and consciousness does not have some other mysterious origin. It too must yield to the rule that everything must have adaptive value: therefore the question becomes, *what is the adaptive value of consciousness?*

My first tentative answer is *modulation* and *optimization* of expensive perceptual hardware. The much-neglected vertebrate cerebellum, sitting at the base of your skull, is probably as miraculous a piece of engineering as life has ever produced: the ability to navigate space and time with uncanny precision is what lies there. Therefore I regard proprioception as the *fundamental sense*, and the one which deserves the most attention – and luckily, the one which seems the most amenable to refinement through meditative practice. How is this possible? Because we are *tool users*: Homo sapiens has the carefully cultivated ability to extend the proprioceptive network around an inanimate object. The tool in the hands of the master is swallowed by the bodily envelope. To the other animals, our hands are magic, because we have the ability to project the proprioceptive field a few inches past the fingers to *incorporate* a tool. That's what a carpenter does with his chisel, that's what a batter does with his bat, that's what a surgeon does with his scalpel, that's what you do with the steering wheel. Meditation as we teach it therefore, is the harnessing of this uniquely hypertrophic neuroplasticity, located especially in our immensely overrepresented hands, to achieve neural modulation: our own perceptual network becomes the raw material, and our apperception the tool. The breath as fulcrum, and apperceptive consciousness as lever: meditation therefore is to exploit human neuroplasticity to achieve refinement of the hypertrophic conscious function, the purpose of which is already a feedback channel, within a living system whose homeostasis is predicated upon various overlapping feedback channels primed at criticality. It is a chain of levers stepping upward in scope – a Rube Goldberg machine of surprising reliability, which requires accuracy, a light touch, and good humor more than anything else.

THE I IS NOT HERE

- 159 -

Consciousness is a localized resonance, achieved via projection of a subset of a richly-connected high-dimensional structure onto a low dimensional space, which when applied as feedback to the generative unconscious network, selectively amplifies its own preimage. Consciousness is a ghost which summons the living.

A conscious topology is a relatively flat and sparse surjective transform, which hides overdetermination and redundancy. Consciousness is not strictly speaking a subspace of unconsciousness, but rather a low dimensional projection which selectively generates resonant activation depending on orientation.

The purpose of this projection-feedback function is the exploitation of nonlinearity in achieving efficiency gains as aftereffect: apperceptive consciousness itself is an obstacle to performance, but the coherence left behind testifies to its role as *calibration*. An animal does not generally become aware of itself except when something goes wrong. A nervous system is potentially very expensive, and must be constantly optimized such that minimal metabolic expenditure achieves maximum result. By applying a kind of "lowpass filter", or radically reductionistic projection of neural activation as feedback, the most salient features can be amplified, retained, and potentially debugged. Most incoming data and the vast majority of the combinatorial possibilities of behavior are discarded: I argue that what we experience as consciousness is partly epiphenomena of this search-and-prune routine, and partly the adaptive exploitation of selective feedback located right at the heart of the issues of intelligence, learning, health, and purposive behavior.

- 160 -

Quite a mouthful – and yet despite the danger we're in, whenever we bow low before this golden calf called "consciousness", my goal here is to find *another vector of attack*: the prejudices surrounding this concept will not melt away just by being ignored, nor can they be "transcended" away – intellect must serve its best purpose and provide the scalpel we need to dissect this unholy thing, this marriage of profundity and sterility, pregnancy and denial. In other words, like our best teachers within the lineage of Western critical theory – Nietzsche, Spinoza, Socrates and Heraclitus come to mind first – we want those same intellectual gymnastics which almost unfailingly lead the human race astray to become the "bent-back bow" which might fire an arrow into the heart of human experience, destroying as much as it creates, setting aflame as much as it fertilizes.

- 161 -

The distilled hypothesis runs like this:

Apperceptive consciousness is a surjective transformation function operating on sensory input, whose purpose is to establish selective resonance with its own preimage, such that the coherence of the unconscious structure is improved. The prime case is always *proprioception*: it is the bodily macrostructure and its need for precise navigation, which necessitates constant calibration.

- 162 -

Looking carefully at the physiology of the cerebellum, I argue that we should conceive of proprioception as a dimensional reprojection of a topologically rendered abstraction of afferent input: the informative density of *nerve endings in space* and *spike trains in time* is radically reduced, translated, and reconstructed within the computational matrix that is the cerebellum.

However, proprioception is generally focused on the thin edge: between the body envelope and potential contact. The arms of the amoeba stretch forth, the cell wall shivers. Apperception therefore is *a kind of interoception* – largely imagined, but built from the ingredients of muscle extension detectors, gastrointestinal signals, and anything on the interior actually innervated, however slightly. Our "inner life" is largely a useful illusion.

The question is whether proprioception as metaphor, or actually as bottomup structural process, can be shown to be constituent of conscious activation. Consciousness is largely a matter of *redundancy*: a consequence of blockage, of signal bottleneck, and therefore an emergent character of many subroutines and their coalescence upon a local domain.

In other words, do we only become aware of ourselves in the "mirror stage", as mirage, as uncanny Doppelgänger, as imposter – or is the bodymind truly something organic, healthy, vital, the real "soul" we've been looking for? My answer is clear enough: the idea that the "self" does not belong here, that it must have some "higher" origin by definition, is only the culminating delusion of a deeply alienated and suspect human creature – it's very common in the last stages of a psychotic breakdown, when it starts to go sour, to conclude that the previous shattering realization "the I is not" should be translated as "I am not here", which quickly becomes "The body is unreal". And that's the history of postaxial philosophy in one sentence. In fact, the circle is coming back around: isn't there a uniquely suspect affinity between the feverish religions of Late Antiquity and their obsession with "higher worlds" and "other origins", and the increasingly common conviction among the commonest people of the 21st century that "the universe is a computer simulation"? I detect the sticky sweet smell of sedentary bodies and emotional stunting.

The apperceptive self must be awoken: the proprioceptive system must be taxed, trained, made to dance. I recently found a movie featuring Fred Astaire at his most brilliant – *The Royal Wedding*: I hadn't realized how much like Bruce Lee his body was – taut, a Swiss caliper of space and time...

- 163 -

Watch Fred warm up in the rehearsal scene on the boat[1]: the first thing he does, is *diminish* his consciousness and *push his apperception down*, into his belly, to internalize the rhythm in his abdominals. The marks of self-consciousness fade from his face. The I is not here.

Consciousness *frames*, while unconscious process does its job: we *do* almost nothing consciously, yet it is present *while* we do things. What's it for? For narration, for gross course correction, for the backpropagating activation of purposive behavior. The locality of consciousness is a boundary ring of *inhibition* surrounding a zone of *selective excitation*: thus it frames what we do, but does not do much of anything, despite our much-cultivated delusions on this point.

- 164 -

Consciousness is a feedback membrane: a containment field, an inner surface, a local refraction index.

Consciousness is localized feedback which pushes the system into higher-order dissipation channels, disrupting local maxima and reified paths, which potentiates the discovery of global maxima. It's fruitful to compare its effect to an annealing technique: consciousness disrupts and excites, effectuating a stochastic search for other dissipation modes.

Consciousness thus has no content of its own, nor any discrete location. It is itself probably an emergent property of every system it "visits".

- 165 -

The amazing thing about the *tapetum lucidum* is not that it reflects light for a second chance at perception, but that it selectively reflects certain frequencies in order to establish a *constructive harmonic* within the eyeball. This "shining tapestry" is a nonlinear amplifier, which ensures greater focus and contrast under low-light conditions: this is a model not only of apperception, but as it turns out, of perception itself. Perception is never passive nor linear, but always grounded on selectively amplifying perturbation of some other living process, such that signal is generated from that noise which biological systems produce not accidentally but on purpose. This is why retinal cells hallucinate wildly in absence of dampening input; this is why a phantom limb springs into place after amputation: perception is well-mapped hallucination.

- 166 -

I've found, both in theory and through years of meditative practice, that the best way to handle the almost radioactive concept of "consciousness" – surrounded by all its flatterers and would-be courtiers – is to treat it as simultaneously *trivial, dangerous, and vital.* It is localized feedback: therefore nothing much on its own. It is generated by anxiety, and thus its cultivation as an end-in-itself is a sign of degeneracy. It is a reparative and calibrating subroutine, and thus its power is almost unlimited.

- 167 -

Consciousness is a diagnostic routine. An animal only becomes aware of itself when something goes wrong: in the cracks between successful instinct, lies the potential for self-awareness. It's a mode of recontextualization, in order to assess danger, futility, failure – *embarrassment* and consciousness are close cousins.

In biological terms, every diagnostic possesses the power of repression. There is a close relationship here to what the ethologists call "sickness behavior": prioritization of healing means the repression of many other functions, and thus is "anxiety" born. Again it's vital not to be caught in the trap of insisting that to "be conscious" is an end-in-itself. Consciousness is a mode of deferral: to disrupt and defer other processes, in order to evaluate the scene and adjust trajectory.

Consciousness is the power of disruption, canceling, hesitation, unbinding. The power of valuation shifts: *revaluation*. Thus its connection with linguistic capacity: symbolic manipulation needs semantic *displacement*, which first happens when instinctual frustration seeks novel paths. Failure engenders frustration, which engenders inventive gratification paths – but it's the element of social witness which engenders *embarrassment* – this is generally why we only see the most social animals as "self-aware".

Becoming aware of something *halts* it – disrupts, interrupts, overloads. It is a testing and debugging mode: thus its connection with the narrative I, dreaming, strategizing, scheming.

Recontextualization: seeing oneself from the outside. The purpose is to assess danger, status, purpose. "What am I doing here?" In the phrase "Who am I?", the key term is not "I", but *who*: this token is an appeal to the context and the group. It asks: "What am I in this place, in this role, with these goals?"

And again, the paradoxical connection between healing and consciousness becomes apparent: recontextualization is the first step of psychotherapy. You must unlearn and undo connections via the power of telling the story again, in a new context, with a novel audience.

- 168 -

Consciousness as diagnostic routine: "What's wrong here?"

Consciousness as selective suppression: "Not now."

Consciousness as confabulatory communication: "What's the story?"

- 169 -

Around 1910, Freud made a critical mistake: despite his better intuitions in *Studien über Hysterie* and *Der Traumdeutung*, he began to assume in his theoretical work that repression "emanated" from the conscious systems, as the first wall of censure against unwanted impressions. But this is false: conscious activity is not nearly powerful enough for something so herculean as systemic repression. When people pretend to forget, when they refuse to know what they know, when they repress beyond recall, conscious activity is at most the distracting magician's hands, the fluttering of a kerchief, the eyelashes of the pretty assistant: consciousness defers, halts, and induces mild hypnosis, but real repression is only sustained through symbolic exclusion, which takes place at deep levels of the unconscious semantic topology.

In fact the situation is almost the reverse: consciousness has the paradoxical power to *lift repression*. Consciousness is a feedback membrane, characterized by a noise complement inducing stochastic resonance, which makes possible the traversal of repressed paths. This is what's known as "free association". Talk therapy is about crossing through repressed semantic zones, in order to disrupt repressed content. But the actual healing work takes place unconsciously.

- 170 -

It's my intuition that there's high redundancy in consciousness. That unconscious phenomena are highly compressed, and thus achieve low redundancy in the message – whereas consciousness is largely constituted by *redundant marking*: how does that coincide with consciousness as a diagnostic and reparative emergent routine of criticality? Coherent harmonic oscillation is redundant, whereas self-similarity is a minimally expensive organizational paradigm: by keeping the system at its critical moment, the system responds like the sympathetic strings of a sitar. It is *poised*, and the slightest perturbation at any scale may affect all scales. This could be the optimal setup because neural computation is *slow but wide*, *shallow* in the sense that the computational steps are few, but so *richly connected* that the total set of possible responses is nearly infinite. The "cortical bit width" is probably unthinkable.

In other words, what we commonly experience as consciousness is largely what's *left behind* by criticality: the famous magic of *unconscious* thinking, as the psychoanalysts know it, is generated by a metastability spawning scale-invariant avalanche – otherwise known as the artistic mood.

- 171 -

It's possible that the kind of feedback and stochastic resonance I've mentioned as key ingredients of apperceptive consciousness, are resultants of this critical metastability: that feedback and resonance occurs not only due to the architecture of neural tissue, but due to the exquisite sensitivity of poised criticality. This implies that apperception is not so much a specialization of complex neural systems, but one of the core functions of neural networks: feedback doesn't have to be conceived as action-reaction loops involving a secondary medium in which to store history – as cybernetics usually does, but rather the effects of a highly resonant scale-invariance, in which perturbations bounce freely from one end to the other, like a tuning fork or a taut drum. What's special about apperceptive consciousness could be only its *localization*, the ring of inhibition which seems to instantiate it: that would tend to create an isolated region of resonance – and perhaps consciousness is merely a result of the *fragmentation* of criticality, of its decomposition and degeneracy, which may have been secondarily developed into a diagnostic and reparative routine in the long history of neural function. In other words, our precious "consciousness" could only be precious because the tree of life turns every error into advantage.

- 172 -

In order to plant the feet of those good-faith readers, those few who stick it out and have learned to trust my poet's errancy to yield analytic rigor, I offer this outline of the theory I'm pitching – as though I were pitching woo on a moonlit night with my bamboo mouth harp outside the hut of my Laotian sweetheart, speaking a secret lover's tongue:

1. Stochastic resonance in the proprioceptive system, and the likelihood of its involvement in apperception. The relationship between consciousness and white noise, pink noise, stochastic filtering and signal approximation.

2. The topology of neural information, or "inner space". Whether neuromorphic experience can be characterized as continuous manifolds, or as discrete graphs, or both. Should we be using signal analysis or graph theory?

3. The geometry of neuroelectrodynamics, or "physiological space". Which can be reframed as: whether and to what degree the ancients were correct in suspecting that exhaustive geometrical configurations constitute the possible stations of the "spirit world".

4. The power of meditation to refine all these processes of consciousness. The degree to which we can reasonably demonstrate an upward percolation of basal neural function into more easily measured intelligence.

I'm mixing waveform analysis in the form of stochastic resonance and thus nonlinear dynamics, with topology and thus graph and set theory. It's a heady brew, which I admit might occasionally sound more fashionable than I'd like – I only happen to think it's correct in this case. I almost cringe when I realize how relevant it is to some of the more adventurous threads in AI research: I too am only a product of my time after all... Maybe it's having spent too much time conceiving of invisible logical paths when writing code, that makes my generation incline so uniformly to graph theory and functional analysis... Space exists but the distance function is mutable. Shifting bits in virtual space will habituate you to think of all experience as "virtual space".

The question is: how can a complex electromagnetic signal be analyzed like a graph, and can a graphical structure yield to harmonic analysis? Even conceived as a narrow signal channel, the number of dimensions involved in a spike train seem sufficiently high to produce topological features of interest: there's the frequency domain, amplitude, the spectral distribution itself, its changing shape over time, its interaction with global parameters or states of excitation, and the variation between signals transmitted over multiple incoming channels. I want to study neuronal activity as music, and oddly it seems the best means of doing so is to reduce it to geometrical and topological relations: relations of ratio, continuity, and path-connectedness. But perhaps that's because music is precisely the intrusion of geometrical constants into the mediation of time: certainly that's what Plato and Nicomachus wanted us to believe. Then is it possible that neural activity is the real "music of the spheres"?

- 173 -

My goal is to employ mathematical language to develop an at least minimally plausible theory of the function and evolutionary value of consciousness, in such a way that also accounts for meditative experience and inspires the practitioner.

- 174 -

Our theory of many-to-one in answer to the question, "what is consciousness?", can now be reformulated:

¬ *Apperceptive* consciousness: stochastic amplification and feedback, which may be a response to signal loss or incoherence. The additive noise so characteristic of "paying attention", both clarifies and suppresses: a ring of inhibition containing a region of reflexive overstimulation.

¬ *Narrative* consciousness: simplification, emergent symbolization, auditory hallucination exploited for producing simulated instinctual coherence for longterm gratification-delayed behaviors. An artifact of human hyperadaptivity: the ability to "think" is largely about the ability to remember and rehearse complicated non-instinctual behaviors. Narration is mnemonic, and thus "thought" originated in song: and so Aboriginal "songlines" take on new meaning...

1. http://bartholomy.ooo/posts/the-i-is-not-here/

WHAT IS CRITICALITY?

- 175 -

There's plenty of evidence that the brain operates at or near criticality. That most of its function which we currently find mysterious, can be described as aftereffects of a nonlinear dynamic system near its critical moments: it's the best theory for describing longrange correlation, which is coordination between physiologically distal and sparsely connected neural tissue producing simultaneous and coherent response, and the mystery of the shallow-but-wide neuromorphic computational style.

That's more or less accepted, or will soon be. What remains to be shown is how or why longrange correlation could come about in general: for example, its known that fractals are generated by those processes which exhibit *nonintegral dimension*, such as the space-filling line, or the coast of infinite jaggedness. It's important to remind ourselves that fractality means it's possible to traverse from a very large scale to a very small scale without a change of navigational function – meaning that one continuous process or *line of force* is capable of traversing not only great distance but *scale*. This is rare to nonexistent in classical physics.

Mathematicians are content to describe the elegant fractality of the process, but don't ask themselves under what condition a system would ever exhibit something like a "space-filling line" – and even those working within chaos theory or synergetics like Hermann Haken, are content to say that longrange correlation occurs at criticality but never seem to ask *why*. The answer didn't occur to me until I began thinking of all these systems as *dissipative*: a pot of boiling water is bubbling because the superfluous molecular energy seeks a dissipative path. Bubbling is the liquid phase pushed to the limit of its dissipative load.

- 176 -

¬ Low temperature systems exhibit high degrees of "order" – but not of the same genera as at criticality. A low energy state is orderly in the sense that it's predictable, redundant, crystalline. It has what they call "correlation length", in that given a point, one can swing a radius in any direction and predict the structure: this is the nature of a crystal and tessellation.

¬ High temperature systems have low degrees of this kind of order, in that one cannot predict the neighboring configuration even given a large sampling of a local area: it has little or no correlation length.

¬ Critical order is very different, as though it were a hybrid.

The insight comes from the fact that chaos should not be defined as "random probability", that's only a *measurement artifact* and the perspective of a technician surveying his data, rather than a theory of dynamics. Chaos and disorder in general is a result of *rapid switching* between competing dissipative modes: the appearance of randomness is a result of high degrees of competition such that each path is fragmented, interrupted, partial. A critical phase transition is the balance of multiple modes holding each other in tension, a *crisis* which is in some ways the realization of chaos. This perfect tension produces its own kind of homogeneity: *critical order*, which has scale invariance as a special consequence of its realization.

It could be conceived as a difference between usable and useless chaos: useless chaos occurs when low temperature order is sufficiently disturbed to produce low predictability but not yet scale invariance; useful chaos is a balance between competing dissipative modes at a poised equilibrium, such that the crisis is fully realized at all scales. This realization is itself an expensive mode of dissipation, expensive enough to contain high energetic throughput.

- 177 -

I provide now a vocabulary list. These are very loose groupings, designed only to help orient the reader to this vast and disparate field. Each term has different origins, with diverse qualitative implications. Some are supersets of the others. Some are technically valid only in very limited scope, some are almost universal. Each one is worth looking into if its meaning is not clear.

Transition:

- Criticality
- Phase transition
- Susceptibility
- Poise

Recurrence:

- Self-similarity
- Scale invariance
- 1/f and pink noise

Dimension:

- Correlation length divergence
- Fractal dimension
- Nonintegral dimension
- Hausdorff dimension
- Critical exponent

Momenta:

- Metastability
- Attractor
- Limit cycle

Event:

- Avalanche
- Cascade

Order:

- Phase
- Order parameter
- Coherence
- Redundancy
- Crystallinity
- Oscillation

Disorder:

- Chaos
- Noise
- Stochastic distribution

Formula:

- Nonlinearity
- Harmonic distortion
- Hysteresis

Amplification:

- Stochastic resonance
- Dither

- 178 -

So what's happening in a phase change? The quantum of throughput energy is straining the dominant dissipative mode, such that it begins to explore transitions to the next mode. At the point that it falls back from the upper mode into the lower, a kind of sideways path is discovered which accepts a larger quantum of energy than it might at first appear. This path is constituted by *scale interaction*: the interaction between scales is the dissipation channel immediately before chaos.

- 179 -

But what is a "phase"? It's a *mode of dissipation*: even if we imagine the phase of any given substance as a highly stable and energetically conservative system, we should think of every configuration as a rich deposit of energy – every form of order as an expression of the potential for change.

This is primarily what seems to hold people back, in the imagination of how and why criticality might be so ubiquitous: a "phase transition" is not some accidental arbitrary point in the graph, it's the maximum dissipation channel available to an overcrowded and constantly perturbed nonlinear system – to be caught between a disintegrating solidity and a freefall one cannot keep up.

- 180 -

What the literature hints at, but does not quite say, is that a fractal arises as a *tessellation of competing dissipative paths*. Like a packing problem in configuration space, the system finds dissipative efficiency via correlation length divergence.

It is as though chaotic fragmentary order were able to accidentally discover this path along the way, as though it's a side-

The Lorenz attractor

channel of nonlinear dynamics: a tessellated nonintegral path, possible only at the junction between phases.

But why not just transition gradually into the next phase? Why a basin? And thus the metaphor of boiling water in an ordinary pot fails, because *constraint* is required: it must be sufficiently difficult to transition to the next order, such that fractal dimension becomes an easier dissipation path than the gaseous state. This pot must be exerting tremendous pressure on that boiling water.

Therefore in neural terms, there must be some powerful constraint at work which forces neural systems into the critical path.

- 181 -

A few caveats:

¬ The essence of a fractal is *not* self-similarity, but fractal dimension: Mandelbrot.

¬ Chaotic systems often follow fractal trajectories known as strange attractors, which show self-similarity, but these are not necessarily caused by criticality: Per Bak.

- 182 -

The strange attractor has *nonintegral dimension*. The usual explanation relies on the trajectory path in statespace. Given a dissipative system with 3 variables, the attractor must be:

1. Less than 3D to allow for *asymptotic approach*: if the attractor had any volume, it would prevent some trajectories from approaching it asymptotically.

2. Greater than 2D to allow for *deterministic neighbor divergence*: if the attractor were a 2D surface, neighboring trajectories could

not diverge without intersection – such intersection would mean a "path choice" and the system would no longer be deterministic.

This is why they talk of "stretching and folding": nearby points diverge, distant points converge, and in this dual action the attractor's shape is drawn.

But this argument is backwards, as most mathematical proofs seem to be. It *argues from definitions* to the plotted data, without attempting any real causal theory. Again I lean toward the *dissipative search*: why would a phase transition produce a fractal in statespace? Under what conditions would this be the path of least resistance?

¬ Why asymptotic approach? Because there is no other optimum dissipation mode. Initial paths will converge here as long the energetic throughput remains steady. An attractor is an order's last compromise with chaos.

¬ Why neighbor divergence? The usual argument from chaos theory says, "because the system displays sensitive dependence on initial conditions": but that's not an explanation of the dynamics. Why should a system be so sensitive to small differences? Because it is *poised*, because it amplifies friction into signal, because the tension between constraint and input energy propagates perturbation upward and downward.

It's important we understand the difference between fractals in statespace, like strange attractors, and a fractal in material space, like a coastline. Yet another thing I have not seen stated in the literature, is this: most likely the fractals Mandelbrot found in nature are the *traces of statespace fractals*. On the other hand, the argument I just gave for the "sensitive dependence on initial conditions" so prominent in chaos theory – the butterfly in Brazil which causes a tornado in Kansas – shows how the realized material structure must in turn shape the possible evolution of the system. In fact, this may be the secret of complex dynamics: that the future evolution of the system and the traces it leaves behind itself shape one another. That is what we mean by "nonlinear".

- 183 -

Correlation length divergence can be conceived as a crystallinity skewed to reside between scales, as though between dimensions. An asymptote creates its own nonintegral dimension.

- 184 -

How does scale invariance arise? As though by withdrawing from each other, each scale expresses every other. In the expansion itself, the pattern is expressed: by running away into higher and lower scales at every scale, it seeks a space for dissipation. The dissipative expenditure is maintenance of this *diagonal of interaction*: the sum of simultaneous horizontal and vertical scale interactions while the fractal is continuously regenerated. Viewed this way, that nonintegral dimension characteristic of a fractal is *the angle of diagonal scale interaction*: the steeper the angle, the closer the fractal is to the integral n+1 dimension.

In this mode, lines of force or *histories of causality* are able to cross vast distances almost simultaneously, because the entirety of the system is held in a kind of semi-crystalline grip, where cause and effect merge. Changes at small scales ripple upward to large, and vice-versa: this is where such high sensitivity to small perturbations occurs, that it begins to resemble "topdown causality", as though there were a center of agency at the helm, a "kubernetes". But this agency is nomadic and ephemeral, being instantiated at any point where avalanche is initiated: and we finally have a viable theory of agency which accounts both for the mystery of "will" and the nonexistence of self.

- 185 -

It's the *fallback* from poised criticality that is so useful: a peak from which hidden valleys can be discovered.

- 186 -

They call it "self-organized criticality", but it's not the self-organization which matters: it's the *exquisite responsiveness* of the whole, its taut yet springy resilience, that responds to small input in a way which is expressive of the whole. Shallow, wide, uncanny: that's the nature of biological intelligence.

- 187 -

The "self" in that formula is only important because there is no locus of control: any given point is capable of avalanche.

The "organizing" is only important because it is the special property of critical order to produce that scale invariance which enables avalanche.

The term "self-organized" is misleading, and expressive of a certain histrionic awe before that which seems to defy our superstitions regarding agency, the mind, the soul – as though we were being generous with nature in this one instance.

An enormous number of inorganic processes can be said to "self-organize": maybe all of physics, maybe everything conceivable in the aftermath of the death of god. And here probably is the root of the matter: "organized" implies "somebody did it".

- 188 -

Most physicists underemphasize the dissipative quality of these systems: criticality is ubiquitous because the universe is so *violent yet monotonous* – or put another way, because energy is superabundant while dissipation channels are limited. The character of young worlds.

- 189 -

Even Mandelbrot never seemed to ask himself what the distinct *design advantage* of fractalic structure might be: like most mathematicians, he's so impressed by the elegance of the encoding scheme that every other consideration fades from view. But that simplicity and rigor might actually be a hindrance in organic terms, since an organism may lose too much adaptive range for whatever savings in encoding is gained: it'd be worth asking the question why life didn't develop further along purely fractalic lines, but seemed to abandon it early on in favor of other symmetries.

It strikes me that the primary advantage of the fractal is not its encoding elegance but its *amplification potential.* Realized fractality acts inevitably as an amplifier of weak and partial signal. Like cracking a whip, the trick is to follow the gentle curve of its unfolding, to induce acceleration without exceeding its accumulating inertia. The cumulative signal grows logarithmically as it proceeds along the fractal scale: amplification response and generative formula coincide.

- 190 -

This assumes that every scaling step, a proportional response grows the signal while matching the antennal sensitivity: such that we don't waste realized surface area for signal transduction. This reminds me of how surfers read waves: they sense which among the series has the best potential to match both the amplifying properties of the shoreline and the ambient perturbations of the medium – if there's too much amplitude early it'll crest before the contours of the seafloor can express themselves.

- 191 -

My guess is that if we plotted the response in a naturally occuring fractal antenna, we'd find it was some proportion of Euler's number, a smooth growth function. The idea is that we don't want turbulence in growth. If we expend too much of the incoming signal – which is nothing but energy – too early in the response, we may fail to explore its full scope. We want a smooth distribution along our growth curve, gaining in amplitude and bandwidth while we eat into our signal budget. From the organic point of view, we don't know the value of any incoming information until it's been fully processed, but some signals are essential, so both efficiency and poised sensitivity are needed: fractal structure seems to allow for the gradual entrainment of some maximum of the network by simply scaling up, from the most responsive smaller scales to the larger less sensitive scales. This should also allow simultaneous signals to compete in parallel. Again during the signal event, neither the duration, nor the amplitude, nor the frequency band, nor the informative value is known until it's traversed.

This assumes that the signal is sufficiently uniform such that fractality has a chance to kick in – to utilize its own self-similar scaling to explore the signal without wasting time nor incoming energy. Imagine what happens when you flick a tuning fork: there's short but intense period of turbulent discharge, in which you've induced useless noise, before enough signal is dissipated that the fork can respond according to its resonant frequency: this is what we want to avoid.

- 192 -

So the question becomes: is this a plausible model for the kind of neural excitation I have in the back of my mind? Namely myoclonus vs the "kundalini shivers":

¬ The kundalini shivers represent the remnant turbulence of an excursion past the critical point, triggered typically by inhalation.

¬ Myoclonic cascade represents the fallback, from some supercritical excitation to subcritical, triggered by exhalation.

It's worth emphasizing that these are merely *symptoms*, not the traversals themselves: these are merely turbulent signatures, vestigial signs. Keeping in mind that most of what happens in the nervous system is *invisible because it's efficient*: the only way to characterize it is via errata, excess noise, essentially *exhaust* – which is why overexcitation via psychedelia and deep meditation is so valuable. We need clues to how the smooth transduction event unfolds, the invisible induction and dissipative loss.

- 193 -

There's a caveat that hasn't yet been expressed: criticality may not yield a grand unified theory of neural activity. It may only represent one possible mode: maybe calibration, maybe rare and expensive peak neural performance, it may not even be particularly adaptive to sustain. My own experience is that it's very brief at best. On the other hand, it does seem that we're supposed to experience critical states every waking cycle, that we transition from sleep states of very low excitation to supercritical positions, that we should be more or less always on the move between these extremes, easily rising and gently falling – that would be the profile of a healthy creature. The larger context here is probably the wake-sleep cycle, and I rarely see this considered in all the talk about neural criticality: that we should be considering a continuum from the broadly coherent oscillations of deepest sleep to critical scale-invariant activation.

- 194 -

So we may call it "diagonal causality": neither fully vertical nor horizontal. This assumes that orthogonal causality is the foundation of Democritean reductionism, which says that the sum of the interaction of similar parts determine the behavior of the whole. "Horizontal" causality is the foundation of Newtonian physics and every account of the result in terms of localized interactions between neighbors sharing the same scale, whereas "vertical" causality is the linear sum of these interactions across all scales. The diagonal causality of scale-invariant critical behavior defies these assumptions.

There's much more work to be done here. It turns out that outlining the psychological resistances involved in "diagonal causality" and nonlinear systems theory in general, requires a revision of the history of science – perhaps nothing less than a review of physics and mathematics since their origin in Babylon. And I do believe the issue is no less enormous: critical emergence challenges the very roots of postaxial Western thinking – or at least their *least Greek* roots.

- 195 -

It's important for me to remember that I ended up in this path because of the need for a *viable mass psychology*. I didn't start with the cute appealing nature of fractals and criticality, and then go look for applications – a la Hermann Haken and Per Bak. I come to this theory because similar vocabulary was growing out of what I learned from COVID – namely what mass psychology is, how topdown causality is possible, what longrange correlation in the social sphere might be. I knew through experience and analysis that it was driven by frustration and alienation – which is nothing but turbulence in social terms: once I understood that such turbulence could create resonance and prepare the ground for an emergent order, I turned to criticality theory in earnest.

- 196 -

I almost wish this theory weren't so attractive: "self-organized criticality", "chaos theory", the "entropic brain" and others like them continue to attract the wrong kind of attention – namely hasty popularization and a reactive dismissal in its wake. But the intuitions persist, and the Zeitgeist won't let it go: the 21st century wants an answer to the question, "what is intelligence?" And it expects an answer in terms of collective dynamics. But much more urgently – and yet I seem to be the only one aware of it as yet – this century begs the question: "what is mass psychology?" The answer to the one is the answer to the other: this is why I'm on this path at all.

BEDROCK

- 197 -

I can only hope this work hasn't succumbed to premature pseudo-scientific posturing. I've asked myself this question more than once, but I always return to what I know through *repeated personal experience*:

¬ That epilepsy lies on the backside of optimal neural performance, and is the risk of those states of high excitation characteristic of profound realization and clairvoyance. Epilepsy is probably the *fallback* from supercritical neural configuration, and not the configuration itself. Migraine is the wicked little sister to epilepsy.

¬ Myoclonus and other forms of neural noise are valuable artifacts which can be developed into healing powers. The "resting state" is characterized by continual noise, which has a sensible distribution. Considering the typical signals in chant and throat singing, this distribution is definitely logarithmic.

¬ Criticality provides by far the best metaphor of intense meditative practice: that both *pressure* and *temperature* are needed in a precise ratio, coincide with the need for simultaneous *containment* and *positive feedback*. Even the relatively simple practice of diaphragmatic control bears these marks.

¬ Proprioceptive refinement is the finest path of meditative discipline, which Qigong revealed to me. Becoming a musician late in life also taught me something important, along with woodcraft and instrument building: vision is tertiary – intelligence reliant on vision is stupid and slow, whereas intelligence reliant on *proprioceptive projection* is fast and brilliant. "Keep your eye on the ball" doesn't mean "visualize the path of the ball", it means "project its path unconsciously": very few understand this difference, but it underlies much of my theory of consciousness and is what makes it distinct.

⁓ Our limbs are phantom. The bodily envelope is a network stretched, fitted, and susceptible to distortion. Many of the esoteric experiences associated with deep meditation and psychedelia are the effects of *decoupling* between spatial resolution and sensory amplification.

These are my guiding lights, strange though they may seem: that strangeness should be witness that behind each is an *unexpected experience* – or more often a long series of experiences stretching back to my earliest memories, which are for me like bedrock.

- 198 -

For example, I learned at an early age how to destabilize spatial resolution. When I was 9, I had a tremendous fever lasting through the night. I hallucinated wildly, I unlearned a belief that things are the way they seem. I saw that a very different world was quite possible, and just as navigable. During that long night, I learned to control the effect such that it became my secret power, a refuge I hesitate to talk about at all because I still do it: it's possible to set your spatial calculation "free", so that distance and scale can be reoriented, along with a sense of bodily ownership. Often the hallucination involves a throbbing in distance calculation, such that space seems to "breathe", and you can eventually gain a little control over it such that objects of intense interest, faces especially, will pulsate rapidly between very near and very far: in fact the distance function seems to be shorted or absent, and only some crude fallback approximation is left. The attending feelings are generally very comfortable, almost womblike.

This is something both Oliver Sacks and Ramachandran talk about openly, but what they don't know is that by remaining in this displaced world, "out of body", one also gets increased access to *unconscious intelligence*. This is part of what I mean by the "dreaming mind": the ability to follow very quiet lines of reasoning, especially those that seem to violate some principle of reality, in order to acquire their intuitive booty. By offloading the overhead of an incessant temporal, spatial, and *narrative* resolution – no one is so "realistic" as the neurotic – the deeper cumulative intelligence of the neural systems are permitted to unfold, and one quickly gains a sense of trust in their strange paths, along with the ability to induce *useful hallucination* at a level largely unknown outside poets and musicians.

- 199 -

The visual hallucination is accompanied by an auditory form: things sound compressed, hasty, lagged – as though a waveform were accumulating at the far end of a sinusoid, crashing all at once. As though I had learned to induce or listen to the *inherent delay* in audio processing: there is a kind of lagging pulse always present, if you pay attention. The smoothness of audio resolution is as much an illusion as the gapless field of vision: what the audial centers actually produce is fairly turbulent and sparse – we guess at most of what we hear. I used to practice with mechanical fans: there is a continual flucuation in relative perceived pitch, following firstly the blood pressure oscillation around the inner ear, but also following some coefficient of attention – amplitude and granularity rises and falls subtly reflecting a continuously updating function of valuation.

In that single night, some synthetic function must have been inhibited, because I found every sensory field opened wide, as though revealed: suddenly I saw mechanics where there had been smooth surfaces, handles and points of inflection where there is normally only a dull "what next", strange delights impossible to communicate.

- 200 -

I'd like this book to be free of the kind of errors and distortions typical of someone who hasn't studied the mathematics seriously enough to handle the terms responsibly. I'd like to plausibly discuss the implications in terms of meditation and spiritual development. We need a revised theory of "consciousness", else we get lost in rechristened abstractions handed to us by the far ancient past: *prana*, *qi*, *dao*, *jhana* and so on were originally borrowed abstractions. At the time, metaphors were drawn from whatever imparted authority: divination, cosmology, sacrificial ritual, and ancestor worship. The *Rig Veda* is the ideal place to study this process, as vocabulary drawn from a sacrifice performed for victorious warriors and worried kings is gradually woven into a priestly cosmology supporting the necessity and power of this ritual, only to be gradually transformed and adopted by the rogue *sadhu*, the lone holy man who rivals the old priest and seeks his empowerment in new signs, namely in asceticism and vigil. He is a "gymnosophist", he wears little or no clothes, he abandons his home and family and the best chances of social success, he wanders and seeks teachings, he gains in independence of mind and attitude. Along the way he finds it expedient to borrow and steal the old vocabulary, putting exciting new twists on each term, while absorbing the language of his fellow ascetics and sometimes inventing his own: this is the process by which a new era supplants the old and yet seems to be its fruition and meaning. We may know we're doing it, but still we must: we need to convince *ourselves* more than anyone, that we're not fools. How does one do this? By communicating our experience successfully: only when another head nods along, do we really become convinced we're not lying. Therefore what choice do we have, but to beg-borrow-and-steal the best vocabulary we can find, to create a minimally plausible theory for the sake of our anxious over-the-shoulder-looking-back, for the sake of the *lonely ape* who is quietly and perpetually terrified by the solitude of modernity and the way our spiritual development only seems to deepen this threat. "Why this way? When so few seem to be headed here?" That lonely ape is there for a reason: because his fears were so often correct – that wandering off into the wilderness for the sake of some crazy dream, some moonlit vision of a gifted but shifted mad primate, too often resulted in the genetic endgame. Your ancestors are the apes that feared the wilderness more than they hated the bickering tribe: that we find it so difficult to think and act independently shouldn't surprise us – the boldest ancestors likely disappeared over the horizon long ago, and it was your grandfather who stayed behind, full of regret and a secret ill-wishing to that freedom. He would rather bicker than learn about himself. He would rather be bored than face a brave new world.

And so spiritual life is an uphill battle – if it's not, you're in a cult. And the cult, precisely as *ersatz tribe*, is what gives birth to the religion, which is what's responsible for catapulting us into the frontier of spiritual life – I would not be here if it weren't for the insanity of evangelical middle-American Christianity, which shaped my inner life from the earliest possible age into a twisted labyrinth. I have been blessed with the old-fashioned Christian conscience, the kind Kierkegaard and Pascal had, the truly brutal one, the one that never lets up, the one that seeks blood in all things, the kind of medieval European dramaturgy that almost no one understands anymore, the kind that had almost as much to do with pagan Europe and its oaths as it did with Israel and the rages of Yahweh. It's this kind of religious training that I find missing in the younger students raised by agnostics and atheists: they don't know how to steer with *dread*, nor anger, and thus not with yearning either. They only want *out*: they know what they *don't* want, but haven't learned to project a fate ahead of themselves with good conscience – perhaps meaning they're less easily fooled by personal hope, but that's a dire problem in the young. A general lack of faith in anything tangible and present haunts 21st century youth, and makes them more susceptible to precisely those collective religious delusions from which European culture just spent so much effort to extricate itself – a new global religion is coming, I can feel it. It's almost as though the Renaissance to late modernity will have been merely a breath midswim: it requires religious torture to craft the kind of soul capable of doing without religious certainty. And if we're not careful, one of us crazy gymosophists is likely to end up canonicalized, snuck in the backdoor like Heraklitus was, right into the heart of the Gospel: one day they'll appropriate our appropriated scientific vocabulary for their own new testament, and words like "algorithm" and "intelligence" and "statespace" will be mumbled over a crowd of cowering peasants.

- 201 -

Am I flexing my intellect or deluding myself and wasting my talents as a psychologist? Shouldn't I be focusing my efforts on sociological questions, historical revisions, philosophical revolutions, spiritual maturation? But there's something undeniably valuable and rewarding in this study of mathematics: it occasionally does function as the kind of soul cleansing that Plato wished for.

Plato, Spinoza, Lacan: psychologists who wasted their time with mathematical fantasy.

Wittgenstein: one of the few mathematicians I would have preferred had taken to psychology.

- 202 -

That a reducing but ambitious, articulate but plausible theory of consciousness helps the student of meditation: that it is designed especially to guide those who experience *profound states of neural excitation* – that my theory makes no sense outside those experiences, that it assumes a familiarity with extremes of systemic coherence, many of which come dangerously close to the pseudoepiletic. That this is something everyone is to some extent familiar with, but no one will talk about: just like talking to yourself, the peripheral ticks and twitches and microhypnoses we learn to cover over with mannerism and mood, are the neglected clues to an entirely new vista. That you are a fragile placidity over a shimmering turbulent flow, that every moment is decomposible into a thousand petulant incipient urges, that the mere act of breathing involves the successive activation of a chain of subsystems and a ripple palpable in the spinal musculature, that when you're doing nothing you are doing a thousand things homogeneously, that sitting meditation can become one of the most intense experiences possible because it seeks to *contain neural excitation until it goes critical*: that's the payoff of all this work with "nonlinear dynamic systems" and the like. By whatever means necessary, we find a way out of the brutality of the Cartesian medical-mechanic: the reigning metaphors concerning the bodymind are infused with some of the most entrenched stupidity the Western tradition has to offer, rooted not in the best of Greek thinking as our mathematics and logic is, but in the *willfully stupid metaphors of the postaxial priest*, especially those of Roman Christianity with its godfather in Augustine, trained at an early age to think slavishly by the Manicheans. Even the Zoroastrian heresy was a betrayal of the ancient fire cult of the Indo-Iranians, clearly the work of a popularizer and an all-too-conscious prophet, one who intuited the needs of the people for a world of black and white, for a more personalized dramaturgy which might explain their anxiety, the first religion addressed to the tribeless individual. That therefore a theory of consciousness is directly implied by the moral revolution of late antiquity and modernity, that I continue to encounter the problem because my psychohistory depends on a working understanding of the nomadic-agricultural tension coming to the surface in peak modernity – which is that point at which nomadic valuations begin to make sense again, in which we all begin to feel like lone wanderers of a wilderness ousted from a fictional tribe, as though we were the crazy ones pushed to the frontier and only occasionally consulted in matters of extreme duress, such as would have only visited a tribe once every few generations...

To be fully modern is to be this tribeless halfmad loner, the one who hallucinates like his ancestors but speaks the language of his peers. As the tribe fragments under stress, it throws out its more exceptional first, in order to sample the neigboring space of innovation: a human community is also a psychosomatic calculator, creeping along strange curves. Modern individuality and its hypertrophic self-consciousness was once an innovation, something marvelled at from a distance and kept away from children. Therefore a "decomposed" unity of experience is not something desirable in itself, but necessary as palative, as remedial education in the basics of subjective elasticity: healthy creatures trip out a lot, without much concern. Nearly every time my dogs nap, they undergo extreme PGO waves during REM, often drifting up to awareness out of what must often be intense dreams – I can see them remark to themselves about the situation, sigh, and go back. The healthy creature is not concerned with transitions of context, nor scale, nor the precise terms of sanity: its joy and lust will take care of pathfinding, and sanity will percolate out from the balance of its ancestral desires. Healthy creatures just find themselves acting for their own benefit, who cares "why".

BEDROCK

PART IV: NARRATION

THINKING IS TALKING TO SOMEONE WHO ISN'T THERE

- 203 -

I frequently speak ill of narrative consciousness and treat it as an obstacle – and this it certainly is. Ceaseless, aimless linguistic hallucination characterizes so much of what it means to be an anxious modern creature. But like every affliction, there is a way to make it an asset. Narrative consciousness must be embraced as a kind of madness: all "thinking" is talking to someone *who is not there*. The first meaning of *Logos* was merely "bundle", "what is gathered", and later it meant "story", and only much later did it mean anything like "logic" or "principle" – in other words, the word itself is a *tall tale*. Our private bouts of reason address a tribe that is not there. The philosopher's path out of this madness has been to talk to ourselves *all the more*: to deepen the conversation, to make it fruitful, to lean into the solipsistic corkscrew until we come out the other end – finally understanding that this madness has nothing to do with us personally, and is common to what it means to be human. We are surrounded by anxious dreamers, who fear they are the only ones who dream with eyes open. So much of what I see when I look into the frightened eyes of the human world, is the fear that only they are afflicted with an incommensurate subjectivity, that it is they alone who must expend so much effort to seem to be normal, that it they alone who seem to stand on the edge of a precipice from which a cacophony of voices emerge, wrapping them in a many-colored coat of desires which they throw back down the hole and disown. But narrative consciousness *is* the favored servant of the pharaoh: the one who can interpret dreams, the one who is always lost within a dream, the one who speaks in riddles as a first language. Right in the center of the most dull and monotonous mental activity, lies the brilliance of unconscious thought. The human world is like a million twitching madmen shaking hands and doing business in the radioactive aftermath of a disaster: they go on as though it weren't there, as though not acknowledging it were essential to the ritual. And yet this human creature is so terribly competent, so fibrous and difficult to defeat: it *will* discover the minimum expenditure to achieve its ends, the least distance path, the least risky strategy. For in the field of petty strategizing, our madness serves us best: no other animal can compete with us in the arts of hysterical displacement, conversion, covert gratification, the return of the repressed, and every other means of achieving social ends while minimizing risk. Why do we think so much? Because we strategize so much. Because narrative consciousness is a *social function*, and the aim of all sociality is hierarchical advantage through symbolic means. Narrative consciousness is inherently linguistic because language has its origin in the symbolic character of sociality. When we talk to ourselves, we are strategizing, we are preparing, we are rehearsing, we are altering memories, we are erasing humiliations, we are talking ourselves back into confidence – we are licking our wounds, gloating, gossiping, and preparing for the next round. And yet this essential loneliness and desperation can be made to serve other purposes: it *can* be bent backwards into a truth function, because the discovery of truth is a creative act – and all creation requires confabulation.

- 204 -

Obviously my own thinking and writing into which I pour so much of who I am, would not exist without this narrative weaponry. Despite all my admonitions against its glorification, clearly it is the *divine madness* which makes this species so interesting. We are infected with symbol: what gives us mastery has also enslaved us. We are the weeping, laughing ape: arguing with voices, afraid of the past, plotting for the future – almost a lost cause. The only liberation is affirmation: if we are to be forever thinking, let us think well. If we are to be forever hearing voices, let it be a good voice. We must find a way out of the labyrinth of voices into the chorus of days: "The grass is a chorus of mundane days, piercing me blade by blade until my body rises."

Narrative consciousness must not be allowed the delusion of control: its proper functioning is to be constantly weaving a tale and leaving a trail of coherent memories behind it. But the illusion that guidance and decision come from this audible voice should not be encouraged: it is the unconscious processes which must be clarified, reified, and amplified – a good story thrives because of its symbolic silences, which grow in the space between the telling and the hearing. The proper function of conscious narratization is *analysis*: it is adept at dividing into parts, loosening, undoing. As a diagnostic routine, it overexcites unconscious symbolic constellation such that novel configurations become possible: this is the meaning of "talk therapy", this is why "just talking about it" can change so much. But the creative and restorative power is not due to conscious process: narrative consciousness is primarily a *critic*. When it praises, beware. When it finds fault, pay attention. When it asks good questions, now it is functioning. The narrative voice which asks probing questions leaves behind it a wake of unconscious coherences like a scalpel slicing through regenerative flesh. Every time we slice, our insight deepens: because the crystallinity left behind, lets us see deeper.

- 205 -

Narrative consciousness likely developed as an excess byproduct of the ability to form complex goal-oriented behaviors: sustained selective repression is necessary for a technologically dependent creature, who must learn most of his skills second-hand and perform them over long periods. The human instinct is to learn complex behaviors through imitation, and narrative consciousness begins as the introjected imitation of social function: children begin their career of "thinking" by talking aloud to themselves as they play, narrating what they do... To think is to play the omniscient narrator.

- 206 -

Narrative consciousness is the strange feedback between the speech centers and the auditory cortex: a crude but effective behavioral organizing function, arising out of the need for highly specialized behavior in absence of instinctual coherence.

Is narrative consciousness an emergent property of linguistic capacity? Is it an inevitable development, once language becomes so overdeveloped? Or is it rather the other way round? Do linguistic faculty and linguistic *dependence* emerge simultaneously from the common problem of *instinctual incoherence*? This is where almost no one understands my evolutionary psychology: the human creature is *helpless* in light of his instinctual weakening, and only becomes strong again due to his delusions.

- 207 -

Let me say it again. It's our talent for *useful hallucination* that makes us indomitable: anyone who's treated schizophrenia with an open mind eventually learns that every bit of crazy is an ingenious solution to the insoluble. This is part of why my work's notoriety has this quality of a *slow-burn*: most cannot, *will not* hear me – the defenses are orders of magnitude stronger than any brew of poetry and rhetoric I can cook. No matter how much beauty I conjure, no matter how many spirits I set dancing to my tune, no matter if I get the pores of the earth to tremble – once a schizophrenic patient told me, in the middle of the most powerful session of my life thus far, that the room was shaking – I cannot defeat those defenses before they're ready to lower on their own: the power was all his, I merely made him feel safe. But if I can make a schizophrenic feel heard, witnessed, and safe, why then all this attack upon you neurotics? It's my instinct: y'all need your assumptions and certainties attacked, from every conceivable angle. Y'all are doing it wrong: liberation and joy lies on the other side of that white-knuckle grip, that mile-wide moat of denial, that ever-present chant of sanity, normalcy, morality, safety, blamelessness, sanity, normalcy, morality...

\- 208 -

We hear voices. But what does that have to do with *being conscious*? Asked another way, how do these voices attract attention? Firstly by being so loud: they overwhelm other systems. Secondly by a long tail of reflection: the feedback between speech and auditory centers leaves traces, stutters, and echoes.

And why do we talk to ourselves? Because we are lonely, and because we are resentful. Narrative consciousness is always *ersatz sociality*: it is either rehearsal, preparation, or memory confabulation. Our famed "reflective reason" is largely rehearsal, largely as-if posturing in front of the mirror of our own witness – and the rest of "immortal reason" is just fabricated *ex post facto*. The logicians are ever on about the same tired gag: they don't want us to know that they sit in smelly closeted rooms having smelly closeted thoughts. They want us to believe in their tired childish fantasies of perfect immateriality and transcendent reason: only so could the reality of their bodily misery be justified – or ignored.

But again is this narrative consciousness, this loud voice, this *cogito*, somehow an emergent criticality of the social communicative function? Is it similarly a diagnostic routine? Resentment gives us the clue: we never talk to ourselves so loudly and compulsively, as when we are boiling with resentment. Dysfunctional social contexts and their symptoms – namely loneliness, alienation, self-loathing – induces a diagnostic hyperactivity within the social systems: thus we hallucinate voices, we "think" constantly, we rehearse, analyze, and prepare. We are spindoctors of a failing tribe, each of us a lost cause in a tribe that never was: the more stressed and abandoned the human creature feels, the louder and more insistent these voices.

- 209 -

The deeply social aspect of narrative consciousness should be emphasized: the narrative *I* is an introjection of the dialogic *You*.

Therefore narrative consciousness retains a special genetic relation to *repression*: talking to oneself is partly a byproduct of the deferral routine every tribal creature full of compromises needs – "La dee da, I'm busy with something I'm not ashamed of, don't mind me". Thus personal narration is not only an instinctual stopgap but a disguise for sneaky ape activities: if you look self-absorbed in an uninteresting chore, you are likely to go unnoticed and thus unharassed. Study the conflict resolution rituals of the gorilla to see it in action: the principal idea of submission behavior is to sedate your superior with his own tribal instincts. Looking thoughtful is to the human what sniffing wieners is to dogs. This also provides another clue why "thinking" usually involves seeing oneself from the outside: contextualization is as important to purposeful behavior, as it is to fooling your wouldbe rivals and persecutors... Paranoia and the capacity for thought are mutually conditioning: it's always the outliers and losers who think the most – something philosophers have not been honest about.

So much of our thinking boils down to this recurrent question: "Who am I in relation to the group?" Status revaluation is a constant habit of nervous apes. The rehearsal of likely future scenarios and the delusional repair of past failures: this is one of those things everyone knows about the human condition but does not address. If you admit how much time you spend arguing with voices, "they" will come and take you away.

- 210 -

Confabulation is an important factor in all "consciousness". But confabulation largely occurs in the realm of narrative episodic memory, *not* sensory coherence: the coherent actionable sensorium has nothing to do with auditory hallucination – in plainer terms, "thinking" rarely makes you smarter – a very old absurdity I must continually underline.

Except perhaps that conscious attention to the sensory whole is part of the diagnostic, and thus paranoia: forming a paranoid transitory *Merkwelt* is the original function of becoming conscious, because it's about addressing a failure in an otherwise efficient unconscious process. "What's wrong here? Who am I?" This is the subtle connective tissue between apperception and narration, and why I permit myself the term "consciousness" at all: both are confabulatory in the sense that reparative contextualization requires fictional stopgaps – white lies are an essential element to functional calibration. Did you follow that or did your egoic defenses prevent it?

- 211 -

That consciousness is a diagnostic routine is surely evident in its somatic apperceptive form, but how does the diagnostic role frame what narrative consciousness does? The idea is that an animal "thinks" protolinguistically when it reevaluates instinctual targeting: *doubt* about the present course is sometimes needed, and thus potentially grounding out and resetting an instinctual process. The principal insight is that instinctual targeting is inherently symbolic – the symbol emerges from the recognition function: thus symbolic reasoning is as old as any instinct, and therefore doubt is as old as both. Next time you see a robin flicking over leaves in a lawn hunting for worms, watch for a moment of misfire and hesitation: that's the birthplace of consciousness.

- 212 -

A word about the famous "thalamocortical loop": its function seems to be to restrict access to the cortex and thereby modulate cortical feedback. Thus has this function attracted attention as a neural correlate to conscious activity: but what's missing, is the insight that the kind of *socially mediated* narrative consciousness induced by the *not merely verbal but written* computer-administered multiple-choice playacting scenarios of most neuropsychology experiments, requires massive inhibition in all other systems to function. "Thinking" like this precludes a great deal else, because it is much more expensive than it seems to us. But most of these researchers have never considered that the kind of thinking they're used to is actually auditory hallucination, and that silently reading printed text involves a minimum of these steps: visual resolution of shapes into recognized words; corresponding stimulation of speech centers such that a simulation of speech is produced while inhibiting the motor actuation; "hearing" this hallucinated voice at the auditory center, which then broadcasts comprehension to many other locales – none of which is *conscious* at all! By the time apperception of the process can take place, all that's left is the echo... Take none of the talk about consciousness seriously, which does not appreciate how little consciousness is able to embrace: the great mystery is not reflexive representation, it is unconscious *presentation*.

- 213 -

Why do we even call this narration "consciousness", if it proceeds so often unconsciously? Because again it's a matter of degree. Most of what the average human being is conscious of, is merely *painfully loud*. There are two ways to wake a doped sleeper: shouting and kicking. In more refined language, we speak of narration and apperception as *redundant marking* and *potential nociception* – again, shouting and kicking! But the crass butt-end of humanity is revealing – not merely comedic relief but comedic revelation:

Dogberry and his nightwatchmen report the truth, if assbackward.

- 214 -

Consciousness is no more mysterious and wonderful than anything else the organism does. That there's so much flattery and low-bowing and pushing-each-other-aside surrounding the palanquin it seems to ride in, is because two primary forces of apenature coincide here: *fear of abandonment* and *vanity*. The immortal soul is dead, the tribe is dead, the nationstate is dying, the nuclear family is dying: what's left to hide our nakedness? What's left as consolation? "Personal identity" is left: and because "consciousness" seems to be the most scientific and unassailable abstraction of this erstwhile soul, it is vigorously and even *viciously* defended – after all my laser-guided diatribe and moral assassination, the only nasty feedback I've received so far was in reaction to my dismissal of that holy cow, the one with "consciousness" painted over its brow.

Consciousness promises *security against annihilation*: *cogito ergo sum* is formulated precisely. The reason consciousness is invoked as *the* answer to the absurd question, "do I exist?", is because the authority of a hallucinated voice is derived from the *social function*: inclusion is existence. If someone is talking to you, you're not yet abandoned. Thus is consciousness portrayed as mysterious, ineffable, and irreducible by people who have no ability to feel such things: it is proof of membership that cannot be taken away. This is the secret behind the modern Buddhist's fearful concern with "no-self": the promise of relief from endless anxiety wrapped in the threat of abandonment – they lurch and grimace around this promise like Kubrick's monkeys around the monolith.

But as soon as the ape is included, it wants more. One of the chief charms of this fickle word "consciousness", is identification with the most impressive feats of humanity: we would like this imaginary unity to be solely responsible for all the technological wonders of civilization. We want the fount of all possible intelligence and the loud compulsive voices we live with, to be one and the same. In fact we feel defensive when confronted with the variety and depth of intelligence all of terrestrial life possesses as birthright: and therefore does a contrary and apesick ape such as myself, feel compelled to preach this *bad news*. But those who know, know it as the best of news: we are not alone in our madness, and while this madness may be incurable, it is also *laughable*...

- 215 -

The famous "unity of consciousness" is a shallow and easily displaced illusion: it's the first piece of unlearning, and something every student of any promise quickly makes peace with. It's not even an illusion: it's a bad habit, a lack of honesty, a linguistic artifact, and an obtuse theoretical assumption which falls apart at the merest touch. Anyone who drones on about consciousness as "the totality of subjective experience" as though they were gazing into a kitschy snowglobe, as though their daily experience were an enchanting bedtime story, as though doubt, loathing, compulsion, and contradiction did not color vast expanses of their actual "subjective experience" – is expressing a *wish*, not an observation. So when they sit down in their pompous, awkward way to *write* about such a thing, they are not *beginning* to tell lies, they are sewing up the edges of a gaudy costume patched together from years of shameful secret skittering and frittering, as though a frantic nocturnal mouse fantasized about spending its days as a fat cat in the sun.

- 216 -

Why is truth so hard to tell? But it's rarely the *first* truth which is so hard: it's the second and third that we fear. Every little defended lie is a dam holding back a flood. This is the reason for deathbed confessions: finally it's possible to unburden oneself without having to face what was really frightening – the cascade of truth, the task of living up to one's honesty.

Do not doubt therefore, that the human creature is highly adept at detecting which denials are essential to their beaverdam – and precisely when and where. The human creature could not be such a skilled unconscious liar if it did not also unconsciously know what is true. Everyone possesses the dowsing rod of truth: the only question is, *what's the social value of truth*? The truth is, not much – hard truths are generally reserved for vicious gossip, the secret ingredient of effective slander, and dirty jokes.

- 217 -

In my first book, *The Moral Disease*, I made the distinction between narrative and apperceptive consciousness. This can be construed as "hearing what you think", and "seeing what you feel". What immediately becomes clear is that there must be an apperceptive aspect to what I call narrative consciousness: most talk to oneself is actually *unconscious*, and it requires years of practice to effectively *listen* to this talk. To be conscious of internal narration is therefore a kind of dialogue, in that this listener also spawns responses – certainly his presence alters the monologue. This is what I mean when I say that you must learn to talk *to yourself* and not someone else. To put this in neuroanatomical terms, this "listener" is most likely constituted by the auditory cortex and its involvement in thalamic feedback, such that the auditory center "listens" to the speech center, while this "silent speech" is maximally exploited to generate affective coherence leading to purposive behavior in absence of the instinctual coherence the rest of animalia still has. In other words, as my good readers will already know, narrative consciousness is a result of frustrated instinct and thus amplified by realized modernity.

And yet it occurs to me that while I'm developing a mathematically inspired language for apperceptive proprioception, I'm content to explain narrative consciousness with abstract entities acting like miniature people in an imagined little room, as is customary among bad psychologists – almost all psychology is merely grotesque puppetry robbed of the bawdy, like a cringing Sunday school play. But why am I suddenly content with sock puppets? The potentially interesting features of narrative consciousness have to do with the semantic relations and corresponding Freudian mechanics inherent in the function of language itself: the proper investigation of the underlying structures with which narrative consciousness must deal, belong to the domain of psychoanalysis. Yet I'm content to leave this "listening" and "talking" relationship in mythopoetic terms, because the process is *in essence mythopoetic*. It is fundamentally a shallow illusion: we are halfmad apes talking to ourselves. That we hear an internal voice only means that we have learned to induce endogenous cheap hallucination as a kind of panacea against the ills of anxiety and frustration. The less anxious and frustrated you are, the more these voices recede, and the more spontaneous and *oracular* thinking becomes.

- 218 -

My most gifted readers will perhaps have already noticed a certain informative recursion in my style. Everything I've said about narrative consciousness has been composed in a single stroke while wandering through my desert wilderness, talking to a handheld recorder: I find I cannot do otherwise if I am to speak of this subject with any honesty. It must be demonstrated *in situ*. Right now I'm standing in the dark looking straight up at the band of the Milky Way poised at the zenith: "surely, in this is a metaphor for the topic" – so says the poet's intuition. Or is it the poet's greed? It *will* be a metaphor, therefore the glory of this spectacle becomes mine, therefore *I* become a mysterious hazy path from horizon to horizon: is it a bridge, or a destination? To that question, the ancients used to answer, *yes*. Just as they would answer concerning the value of our hallucinated voices. Is our thinking the prelude to madness or the means to wisdom? But just as with everything that lives, just as the trembling whirling of every galaxy demonstrates, one cannot *be* anything except by continually *becoming* it.

WALKING AND THE ART OF EMOTIONAL MATURATION

- 219 -

One of the practices which has kept me sane and healthy over the years, is my habit of taking long walks. Even when living in some of the densest cityscapes on the planet, I have always found it possible to navigate my way toward a long, striding, refreshing march.

The human being was made to walk: our upright skeleton, our svelte musculature, our plantigrade feet, our bobbing oversized head all need to be exercised on long walks to make sense again. Without a daily walk, preferably through a pathless wild, I begin to lose my sanity, my sense of humor, and my good night's sleep.

Yet I believe that many of you don't know this. Much of the modern firstworld struggles with "exercise" as though it were something exotic and far away. There is no need to jog, no need for equipment, no need to get in your car: just use those legs and go somewhere, for the earth is large and round.

Moreover, one of the secrets of walking as opposed to all other forms of exercise, is that it *tunes the mind*. The human mind responds to a walking pace more than any other: just as 90 beats per minute is the central point around which human music revolves due to the temporal range of the human heartbeat, so is the best of human thinking centered around nearly the same rhythm: they call it an *andantino* for good reason.

Therefore I will outline for y'all, my approach to *the maximally productive walk*.

- 220 -

- Go alone.

- Leave your phone and all other distractions behind.

- Establish a good rhythm and keep to it.

- Challenge yourself to go a little further than you think you want to.

- Early on, let your mind wander. Allow yourself to become chatty and distracted. If you find yourself talking aloud, don't attempt to prevent it. However, don't allow the thread to become a mere indulgent fantasy: you must really talk *to yourself*, not someone else. You must learn to take yourself aside, as it were, and consult with your soul: treat it respectfully but frankly, like a good friend.

- Use free association during the early stage. This will allow you to discover what it is you need to talk about: once you discover a theme or a stronger emotional response, focus on this issue. Trace the difficulties, look for knots. Treat your emotional responses like tense musculature in need of massage.

- 221 -

Some of you may discover what seems like an endless need, or an inconsolable loss, or more reasons to be depressed. Many of you are already running from yourselves, and therefore cannot afford this kind of reflection: however, it's possible to use walking as your grounding, your emotional safety, even your justification. The logic of depression says: "You're hopeless, you've already lost, there's nothing you can do." But something as simple as a long walk in the fresh air can weaken this logic and make it seem pale and distant. If necessary, trick yourself: begin with a short walk, and see if you want more.

The simplicity and centrality of walking lends itself well to clearing out bullshit: even when nothing seems to make sense, even when you feel out of control or powerless, you *do* know how to walk. In this practice, there are times when your feelings will spiral and bubble, swirl and mock you: use the simplicity and reality of each stride to straighten and strengthen your attitude. Feel the solidity of the earth, the competence of your feet, the power of your heartbeat, the inevitability of your breath: these are gifts a billion years old and more, and they cannot be defeated by the breezes which wash away the sandcastles of your merely conscious mind. Consciousness is almost nothing, a mere plaything, mere epiphenomena, upon the great sea of your body and the biological forces which created your body. You are in fact the representative of your ancestors and a rivulet of the torrent of life reaching back to the beginning: you are related to everything which lives. You can never be alone, and we are each of us alone forever. This is the power of walking upon the earth as a native creature.

- 222 -

This practice is part of the burden and gift of solitude. Walking in this way is an excellent supplement for those beginning a meditation discipline – but it is no substitute. You must familiarize yourself with how loud and urgent the voices in your head really are: do not be ashamed, do not lie to yourself about it. You will not begin to quiet that noise until you learn to give each voice space, and deal with them one at a time.

- 223 -

The earth is everywhere and always with us. Even the dirtiest and narrowest alleyway is planet earth. Most of the earth is still thankfully "the outdoors": do not lose sight of this.

- 224 -

Do not indulge the insecurity which wants to fuss with "gear": this is only another means of preventing difficult feelings. Every year I see tourists pass through my desert wilderness burdened with a thousand toys, a hundred machinations, and grandiose expectations – but not once the humility to get out of the car, sit on the ground, and listen. Most of the "recreational" activity aiming at nature, seeks to *obliterate* the experience of nature: to make it into another tiring social event, another falsified story, another pointless toil, another lawn to mow.

Approach nature with humility and simplicity, and she will speak softly to you. She will gradually change you back into what you are supposed to be.

- 225 -

It may sound strange, but walking can be a form of *spiritual activism*. Walking is an excellent means of bending the rules and changing expectations. It is also a way of discovering how free you are and are not: can you walk through this guarded parking lot? What about this strange little concrete tract of riverfront? Will someone in this upper middle class gated community call the cops if you take a long stroll? If you walk through this lively neighborhood, will the racial tension be too much for you? Is it you or them who's tense? What happens if you say hello?

Manufactured suburbia is the worst in this regard: there are millions of acres of prime North American farmland paved over with tract housing, in which the emotional repression is so tightly controlled that even walking on the sidewalk seems to constitute a misdemeanor. The tyranny of the automobile wants to annihilate simple freedoms. I watch people get into and out of their car like they were bowing in prayer: it is a ritual they love, because the automobile is a surrogate body. They love the moment of climbing in most, but they also enjoy the fussy self-importance of parking and getting out – but once they're in it, their repressed feelings surge up again and so they must drive hastily. Once they're out, the reality of the body oppresses them and they must hasten to cover over it again. The cycle of meaningless toil is not accidental.

- 226 -

The human spirit loves and needs stories. Every feeling is a little story, with a beginning, a middle, and an end which hints at a new beginning. You must learn to savor your feelings like a sommelier. Every feeling has a right to exist no matter how earthy, bitter, and forbidden. To draw out this innocence and develop a freedom from stagnant expectation, there are two interrelated methods: talking and walking.

Talk therapy works because language has special leverage which reaches deep into unconscious structure. As Lacan says, "the unconscious is structured like a language": this also means it's manipulable like a language, like a multidimensional spiderweb, like an instrument with a hundred million strings...

Walking is the muscular parallel. As my old analyst used to tell me, "Say the next thing". In order to give your feelings this right to exist, *keep the body in motion*. Keeping the body in motion gently stirs the pool of the mind. Therefore while much of my technique is derived from psychoanalysis and the stargazing posture of lying flat on your back, wondering out loud to an invisible presence – there is an increasingly popular kind of conciliatory indulgence and half-therapized attitude toward emotional landscapes which would threaten to make what I'm saying into just another truism for those who want to mellow the human spirit into a cloudy kiddy pool at the spoilt brat's birthday party, with too many presents and no genuine friends...

But I'm talking about evolving your feelings, about forcing them to develop into something actionable. I'm talking about finding a path of maturation despite all obstacles. I'm talking about *doing something about your life*. I'm talking about laying the foundation for a lifelong friendship with yourself.

- 227 -

Every walk should be a little story. Every landmark can be used as a point of association and thus a metaphor for your progression. If you stumble, pay attention to what you were thinking and feeling right before it happened: it was no accident, but an act of self-sabotage – why? If you find yourself going the wrong way, ask yourself why. What does this way mean to you?

You can use distant landmarks and difficult passages as a limiting point for any one thread. By giving yourself a limited domain in which to come to a minor conclusion, you stand a better chance of fooling yourself into doing so. Yet in all of this, one should not try too hard, and one should always allow the unconscious mind to take the lead. Don't anticipate answers, find them. Strive to be honest with yourself: honesty does not come all at once in a moment, it is a long gradation and a lifelong task.

- 228 -

Inclement weather is no excuse. I love winter hiking. Your heart, your lungs, and your circulation will meet the challenge. One of the benefits of hiking through a snowstorm, is not just that you'll have a chance to improve endurance – but that it offers the opportunity to expand your range of *emotional control*. Walking alone in a gloaming wilderness through a blizzard, will provide a concrete opportunity to expand the set of feelings which you can handle without panic. A little brush with mortality will make your petty anxieties seem - petty.

If you practice consistently, you'll soon be walking many miles daily without straining yourself, and you may look back and wonder how you ever could have been so weak.

- 229 -

Although it's important to assert that you already know how to walk, in another sense many of you have no idea what you're doing. Some of you have never learned to walk over uneven ground – there are many adults who consider themselves fully competent and even the pinnacle of creation, perhaps even an example of universal aspiration itself, who cannot walk over anything but a perfectly graded, dry, paved surface: at which point the arrogance of civilized humanity looks more like a Mr. Bean skit than anything else.

If you want to know what masterly movement looks like, watch a film with Toshiro Mifune. He studied Aikido and it shows. He always moves with his knees wide, his weight low, his feet firmly planted, and his lower back activated. Or study Cary Grant: he was an acrobat, and it contributed greatly to that ineffable grace.

Do not take walking, or standing, or sitting for granted: these too can contribute to an athleticism, which when properly developed has the power to fundamentally change the experience of your body. It's in the small things that power over yourself lies hidden.

INEXHAUSTIBLE WELLS OF SADNESS

\- 230 -

If the reader would like to experience a more visceral representation of the power of narrative consciousness and reflexive dialogue, may I recommend the Indian vocal tradition. There are threads running through Carnatic Indian music reaching back unbroken to the original African migrations: listen to the ghatam to hear them. But it's in the northern vocalists that the magic of dialogue reaches its full pitch: in the *Khyal* schools especially, they talk to their feelings, they build them as slowly as possible, they work circularly to build an unassailable emotional foundation from which to rise. Jasraj or Bhimsen Joshi can demonstrate mastery of the endogenous processes of emotion: invocation, gentle repetition, a swaying variance, the willingness to allow the impact of silence, the maturity not to need to show off at every opportunity, the patience to allow the feeling to develop and build in its own time, the assurance that inspiration will come... Everything a good musician learns, demonstrated over the course of a structured thirty minute improvisation.

\- 231 -

I once had the good fortune to see Mohammed Reza Lotfi in concert. There is a long history of mutual influence such that the Khyal and Sufi schools are not so distant in aims and means: this is not just "holy" music, as though we were to feel nothing but our *vain wishfulness*. This tradition exploits the inexhaustible wells of *sadness* which inevitably accumulate in old civilizations, not just as spice and roundness, but a depth from which to look up: as though peering through an accumulated haze of futility, the stars are allowed to burn ever brighter and more passionately. It's as though they know how to stretch the poles between every feeling, to allow them breath and breadth, to open the pores of the immediate moment and show you something new right where you have overlooked a thousand times. That, my friends, is also the gift of meditative discipline.

- 232 -

Want to hear something strange? When I look for an analogue in Western music, what I think of isn't hiphop but the best *heavy metal*: that throbbing, all-consuming, all-embracing, all-annihilating rage, which seeks its redemptive ecstasy in violence and terror. I saw *High on Fire* in concert once, and as strange as it may seem to compare a screaming, stoned, sweat-drenched thrasher to a swaying Sufi reciting Hafez like a cosmic lullaby, I see a similarity in the approach to the *emotional problem*: how do we feel every bit of what we need to feel, lose our expectations of outcome, yet know exactly where we are at every moment such that we always *stick the landing*?

THE BURNING BUSH

\- 233 -

Most of my best thinking occurs as I walk. The first mile serves to clear my head. At this phase the voices swirl and tumble over one another, while the rhythm of the march settles the dust and the pertinent questions rise, precipitate and find voice. I never force it, I allow it. I sweep the ground of this "soul" like a front porch in perpetual autumn. By the time my legs are tired, my lips are usually moving. Walking is my secret weapon. It's what makes my prose seem so ambling, laissez-faire, and yet unaccountably *dogged*: tenacious in its twists, with its ability to turn a ragged edge into a pulled thread. What's disorienting to the average reader, is the number of concurrent threads entwining my digits, the way I seem to allow the discourse to become a tangled ball only to give it a tug and reveal that there was no knot. It's a high thread count we ply. Who knew tenderness could be obtained from density? As though by accounting for every stray thought in the midst of articulation we impart the sense that sudden interpretative crystallizations are safe in this place, because in the midst of all that diversity, foreclosing finality is not possible. Sometimes I feel like I'm running an orphanage for human experience. – As a child I had the idea to become a *garbage man*: the way they hung off the back of that truck with one hand looked like my meat and potatoes. But it goes deeper: I feel an advocacy for everything denied, repressed, *verworfen*. Growing up in an agricultural heartland, there was an enormous sugarbeet processing plant near my childhood home, the odor of which everyone claimed to abhor. But as a small boy I made the counterclaim that it was not merely a "bad" smell but interesting – because they weren't experiencing it, merely repeating one another. Something else comes to mind: amidst all the fascinating rust involved in 20th century industrialized farming, there was a very curious property I would pass on my way to school every day which captured my imagination. Most observers dismissed it as just another junkyard full of worthless rotting relics, a common sight in the rural West. But it sparked in me a vision of freedom that's never left me, because in that yard there was not only a tractor, an old truck, but a gleaming aluminum airplane, which to me revealed the possibility of endless creativity, invention, repurposing. To be a Da Vinci of the discarded: that dream has persisted and is reflected in my lifestyle as a stubbornly offgrid redneck, and what must seem like an equally inexplicable advocacy of the repressed and forbidden content of our various worlds. I don't usually talk about the concrete details of my current life, firstly because they are a precious refuge, and secondly because they invariably induce a volatile cocktail of envy, disbelief, and dismay. "Why would someone capable of our kind of success choose to neglect it, unless there was something wrong with him? If I'm not permitted escape, nor shall anyone." First issuance of ape law: *what cannot be possessed must be destroyed*. Therefore I don't attempt to share the laughlines of my face, the austerity of my desert home, nor its peace. Nor would it do any good to explain that I have of course escaped very little. That my choices add up to a *cumulative confrontation*: with myself, with my flaws, with my past, and with this continually cresting history with as much clarity as I can stand. What I do attempt to share, is the little handheld obsidian mirror I carry, the scrying stone which has the ability to reflect the neglected aspects of *your experience* – everything lying in wait, lying in rubble, perched on loose scree, and behind locked doors, velvet ropes and buried under heaps of garbage. I come bearing gifts – maybe still a little encrusted with filth and stink - but if I hand you another polished product, what good will it do you? You're surrounded by spit and polish, swimming in the spittle of slick ploys, and just because the chute you're sliding down ass-first has walls, you imagine it must end up somewhere safe. But if I leave a little of the work to you, to clean and set those gems or leave them in the rough, gazing on their potential, then they have a chance of becoming yours, not merely another dead artifact but a living power object. I want to teach my readers to build their own *instrument of insight*: something that stirs the suffering fragments, provokes them into a whirlwind, so that the vortex may speak. The "burning bush" is what we used to say – now we talk about the "unconscious coherence of repressed ideation", but the sense of liminal apoplexy remains.

DREAMWALKER

- 234 -

The highest tier of psychologist is the *dreamwalker*. He learns it not in sleep, but in waking: the ability to capture, widen, and sustain those fragments of dreaming which occur at almost all moments of sentience. The art of *microhypnosis*: he does not so much "enter" a dream, but *dilates* a fragmented dream opportunistically with unconscious intent. This is why the most effective, and also most dangerous psychologists are *instinctual*: one has to know what one wants. It's not possible to generate deep communicative resonance with merely conscious activity: most meaningful transmission occurs in omission, association, subtext.

But what seems like the ability to hypnotize, is largely the ability to *cancel distraction*: actually every sentient creature moves between many simultaneous and orthogonal trances at once, the interoperations of which produce contextually rich behaviors... Every trance is the result of the isolation of a fewer set of perceptual-behavioral loops than normally operative. What then causes that *stasis* which gives it away? Just this: the temporary dominance of one instinctual systemic response over many others – instinct spends most of its time waiting, and defaults to off. Think of the stalking cat: its eyes fixed, its musculature poised, all other responses suppressed.

To "mesmerize" is to produce a potently multivalent signal while *deferring interpretation*: to keep everyone awake, interested, but suspended between stimulus and interpretive response. Nothing breaks beautiful trance like premature interpretation: thus the droning, repetitive, uninformative talk of most modern folk, which manages to combine suspiciously optimistic monotony with a marching thrum of numbing certainty at every step. This is *antitrance*: a shallow fitful sleep which rests nowhere, which constantly strums the same note in response to every input in order to preclude associative traversal... *To suppress premature interpretation*: that happens to be one of the ingredients which Freud discovered was essential to the exploration of unconscious valence – everyone likes to forget that Freud began as a neurologist exploring the power of hypnosis.

If you get the feeling sometimes, that my writing induces mild trance, you are beginning to understand me. But no one can induce receptivity where there is none, and *every* dream is a "shared dream" – because we are never alone within this sloshing nucleic acid tidepool called terrestrial life: to be conscious is not to *be* alone, only to produce the illusion of useful unity where there is plurality. It's no accident that every nomadic people on the planet considered the dreamstate to be that place where contact with ancestors was most likely... So what are we doing together then? I am speaking to you across the narrow genetic footbridge which separates us, a mere 200,000 years of branching at most: as a single species, we are *of the same body*. Our nervous systems resonate so readily, because they are hardly different – only I know how to make mine sing with words, such that it becomes a swaying rhythm built on the rhythms of heart, blood, stomach. But it's *your* power I'm using: all I can do is defer the distraction of your garrulous narrative consciousness with a flood of irresistible poetry, such that latent wakefulness finds itself in the midst of a suddenly stilled landscape – depth of dreaming comes when the *wakeful* coexists with unconscious topology without disturbing its dendritic overdetermined process... In other words: the ability to witness, respond, and even manipulate, without imposing the brutality of singular valence so typical of the forebrain. The dilation of fragmented dreams: there is no more significant channel of communication I know of, and I've looked.

AFTERWORD

- 235 -

There is no chance this theory is complete, and an exhaustive theory should not be our goal. Philosophy, psychology, and science has always *regressed* precisely to the degree that it attempts the "final word". Our knowledge deepens when we embrace the incomplete. Mastery is not about knowing what and where and why, but the *refinement of method*: not the possession of answers, but the regenerative power of questions. *We have no idea what consciousness is*: this is the point of beginning and the terminus of the inquiry as well. What are we left with? A lexicon, a set of memorable metaphors, a practiced musculature, a small map of the available perspectives, and an *indelible curiosity*: if we have become more likely to generate insight in our stumbling, then we succeeded. But every insight is transitory, and fades as the vista shifts: there's always another obscuring mountain, another unconsidered dimension, another younger voice. Let the leading voice be the youngest, and the eye the oldest. There is more experience than can be voiced: which also means, there is always more articulable ignorance upon which wisdom may feed.

- 236 -

And now I've done what I previously refused to do. In some ways I've written this book against my better judgment: isn't this prescribing sweets for a belly-ache? Haven't I arrived to the marketplace with yet another cartload of trinkets for the fop? More fashionable means of intellectual deferral for the ambitious but confused – more candy where bonebroth is needed. I'd like to think that my theories edify the real path: health, humility, a return to the blessed untamed earth. But one will always be used and abused: to be willfully misunderstood betrays that one has been understood all too well.

The ape will have his fictions. He needs his *meta-physis*: an imaginary plane of causality, an explanatory myth in place of the freefall of awakening. He must contain his anxiety, he must give shape to his dreams in ways that spoil their depth and insight, he must invent a way to be *sure*: the ape must be confident in dealing with abstractions that would drive any other animal unthinkably mad. We live in a world more absurd and meaningless than ever, more mechanical, alien, and bureaucratic than ever, more gray, indoor, and pale than ever: this poor creature wants sunshine and movement, challenge and reward, destiny and belonging – but doesn't know how to get it except with *yet more sedentary thinking*.

And so I'm angling to be a *fisher of men*: everything the path will eventually take away, can be clothed in an attractive hypothesis without violating our intellectual conscience. At least in flyfishing, the difficulty of the art is sufficiently steep that it ennobles the deception: no longer merely bait, but a mimicry that brings predator and prey together. I've become more like them, in the effort to reach them: these theories are not themselves a path of liberation, but the *communicative relationship* they represent is. In turning around, after so many years of discarding and discrediting everything which wanted to tell me what meditation and consciousness is, after learning to do without a theory, I find the value of this work not in the crystalline result but the *deeper appreciation of the mess*: constructing a barge for this swamp is even harder than it looks. I sometimes feel like Huckleberry Finn losing Jim in the nighttime Mississippi fog: isn't it amazing how defended and divisive our pet concepts are? How quickly liberation and self-realization becomes gaslighting, dogma, betrayal of good sense? "Is I me, or who is I?"

- 237 -

> My lonely dark is populated by many a guiding daimon, whose *tapetum lucidum* I depend upon.

What is a *tapetum lucidum*? It's a thin membrane which gives light a second chance, which collates scattered fragments and makes them serve as resonance, which illuminates the eyeball like a Chinese lantern, turning dim noise into amplification. This is a metaphor not only for consciousness, but this little labor of love you hold in your hands. A handful of lumens, a couple hundred more spike trains, a milligram of serotonin, a more lubricated synapse, a more poised attitude, a slightly quieter ego, a mouth more ready to smile, a more indelible glint in the eye. If all this mathematical vocabulary and my furious brushstrokes add up to a little more endogenous insight in even half a dozen students, I will have succeeded. Genuine meditative discipline has always been passed on very sparingly, and not just so that the secrets are preserved – but because the alchemy is so delicate, requiring a profound stubbornness and undamaged determination paired with a kind of desperation and thirst for excellence rarely seen in those who choose such esoteric paths. In other words, a kind of self-benighted and willing loser who's wandered into spiritual practice despite being very capable of success elsewhere, someone deeply confused about his worth, seeking to prove it in an arena where almost no one will be able to verify it. Meditation is difficult but entirely possible to master – what is almost impossible to achieve is its transmission and any kind of social reward for having achieved it. Thus is its refinement so rare. Therefore may this book fall into the hands of the confused and gifted. May it light up the shores of a rocky youth. May my work reflect a little flame into the choppy waters of spiritual development. May I offer a few more lumens, that very candlepower I stared into, down to its last thread of wick: that same wick which plunges through your waxy frame, sitting upright in your own lonely passionate practice, stacking your perseverance on top of mine, threaded together by this thin fiber of insight.

Fool's Bluff, seven days before summer solstice, 2024

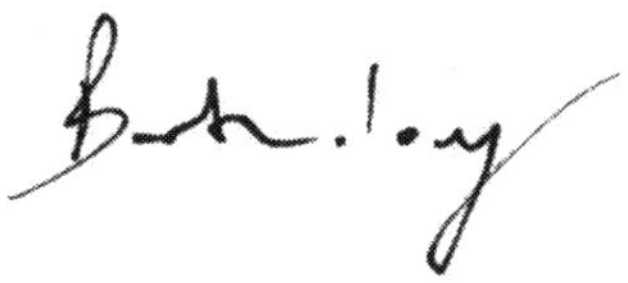

Made in the USA
Middletown, DE
27 September 2024

61593217R00151